The Complete Guide to High-End Audio

John Watkiss

July 28th, 1961–January 20th, 2017

Extraordinary Artist,
Exceptional Man

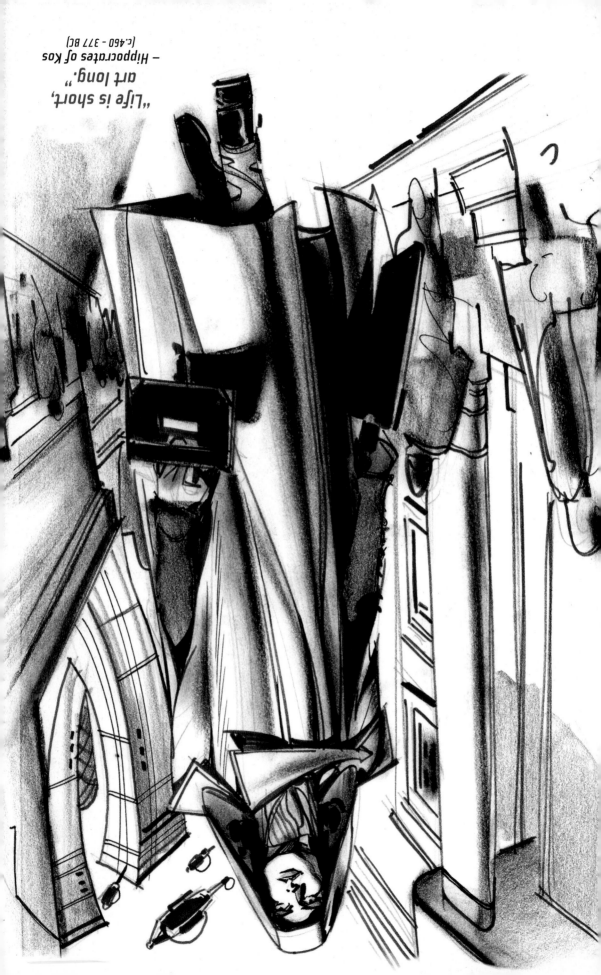

"Life is short,
art long."
– Hippocrates of Kos
(c.460 - 377 BC)

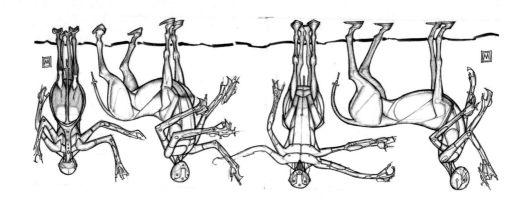

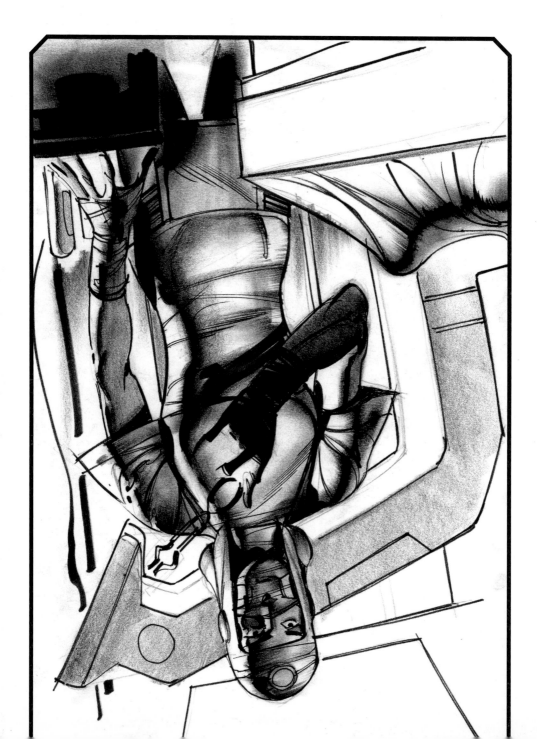

John drew over 100 amazing pieces of development art for the series. These are some of our favorites.

Hello, everyone

and thank you for reading the first storyline of SURGEON X. I've been telling stories in other mediums for years, but this is the first comic I've written, so I thought I'd tell you a bit about how SURGEON X came to be.

For the past 16 years I've worked as a TV Producer/Director in documentaries, factual drama and animation. I've always been fascinated by stories of life and death and of science heroes and villains. I had an idea for a story about a female surgeon working in the near future, in the midst of an antibiotic apocalypse. To me the obvious medium for the way I wanted to tell this story was a comic. How was I going to get this made?

I applied for a grant from Wellcome Trust who are a British charitable foundation that mostly fund science research, but they also believe in public engagement. In November 2014, I found out we got the funding. The cool thing is that Wellcome Trust are like a modern day artists' patron. They have no editorial interference; you get total creative control – a dream! They also encourage you to take creative risks, invent and innovate. I'm grateful that such an organisation exists.

I've read comics all my life, but this is the first comic I've written, so I took a wild shot in the dark, and approached Vertigo founding editor, Karen Berger, and to my delight she responded positively to the idea. During the last two years, Karen has been an inspirational editor. Patient, insightful and generous of spirit - it's been amazing to work with someone of such calibre. As a storyteller, I have a newfound respect for comics creators – it really is an underrated art form, which should be enjoyed by a far wider audience.

Karen introduced me to the incredibly talented John Watkiss. Our team expanded to include more brilliant comics talent - colourist James Devlin, letterer Jared Fletcher, designer Richard Bruning and artist Warren Pleece who helped out on the last issue, when John became very sick, and sadly died.

John had an incredible range as an artist having worked in film, comics, animation, as a teacher of anatomy at the Royal College of Art, and also as a fine artist. Our favourite meeting spot was the British Film Institute café at the Southbank on the River Thames. We would meet to discuss the comic, but inevitably ended up talking about philoso- phy, art, science and we spoke extensively about his life as an artist.

Even when ill, John was determined to finish SURGEON X. And he did – he completed all the roughs and all but ten pages of the inks. His belief in this project humbles me and I'm devastated that he's gone. He was a unique talent and we miss him very much.

We worked hard to get the medical aspects right because it makes the storyworld more layered and unexpected. We had access to an incredible team of science and medical human- ities experts who not only made the stories more authentic, but also sent me off-map, a dream for a writer.

What else? SURGEON X is an APP, too. Having worked in film and animation, I also wanted to use those skills to expand on the storyworld. We couldn't find a comic reader that could support film, animation and audio, so we had to build our very own.

The project has taken two years to complete – with over 175 pages of comic story, 100+ stunning development drawings, 70 minutes of film documentary, 25 minutes of animation, 50 minutes of specially composed music, social media content and a website. I hope that over time fans of SURGEON X will explore this material via the app and online.

The antibiotic apocalypse hasn't happened yet. How bad it gets is not just up to governments or doctors - it's up to us all. We need to apply pressure to our respective governments to intensify antibiotic research, develop better diagnostics and reduce the use of antibiotics in farming. We also need to listen to our doctors if they tell us we don't need antibiotics. In a way, we're all complicit.

As much as I love telling this story and moulding these characters, here's hoping that in 20 years' time, the world of SURGEON X won't become our actual reality. But, only time will tell...

Sara Kenney

Sara Kenney

@WowbaggerUK
@surgeonxcomic
www.facebook.com/surgeonxcomic/
www.surgeonx.co.uk

The app is available from the App Store and Google Play

WHO'S WHO

SARA KENNEY

is a writer/producer/director based in London, England. She started her career working in science, but left a good job with great career prospects to work in TV. Sara spent 4-years at the BBC, but in 2004 left to go freelance working as a filmmaker on documentaries, drama and animation. SURGEON X is Sara's first comic book.

JOHN WATKISS

was an artist who had worked in both comics and film. He began his career in London as a portrait painter and illustrator and then also taught anatomy and fine art at the Royal College of Art. He had worked for Derek Jarman, Saatchi & Saatchi, Ridley Scott Associates, Francis Ford Coppola, DreamWorks, 20th Century Fox, and Disney. In comics, Watkiss' work included *Sandman, Conan, Deadman* and *Ring of Roses*.

KAREN BERGER

is the award-winning editor and founder of Vertigo, the creator-driven imprint of DC Comics. She led Vertigo for 20 years, publishing over 300 properties, including the acclaimed and bestselling series *Sandman, V for Vendetta, Preacher, Swamp Thing, Fables, Hellblazer, Y–the Last Man* and *100 Bullets*. She is now the editor of Berger Books, a new line of creator-owned comics and graphic novels from Dark Horse.

JAMES DEVLIN

is a colourist from Glasgow who has worked on various comic titles including Vertigo's *Testament* and DC's *Supergirl*. Along with SURGEON X, he is currently drawing and colouring the grindhouse horror-action-thriller *Vietnam Zombie Holocaust*.

JARED K. FLETCHER

began working at DC Comics as part of their new in-house lettering department after graduating from the renowned Kubert School. A few years later, he left to pursue his freelance career as the proprietor of Studio Fantabulous.

RICHARD BRUNING

has spent a lifetime in the comics business, both as a creator and as DC Comics' Creative Director. An award-winning graphic designer, he's created scores of logos, including ones for the Vertigo imprint and the San Diego Comic-Con.

SURGEON X ™

Or perhaps the best.
When I think of my sister...

Or is it that I know what comes next?
Millions could die, and this disease
will bring out the worst in people.

"...Because I feel no
remorse for killing Vikram.
What's wrong with me?
Am I in shock?

Can I trust him?
Can I trust Dad?
I don't even know if
I can trust myself..."

But maybe he just
sees things more clearly
than the rest of us.

ring

Rosa, you have got the worst bedside manner *ever!*

Pull over here, Caspar.

Shit-- one thing after another with this antibiotic crisis.

Drop me at Victoria. Home's easy from--

You're so much *smarter* than that crappy line.

Sorry, Caspar, I'm just in a weird place. Can we keep it *casual* and see what happens?

Really? You know I've liked you for a while, Rosa.

But you're doing my *head* in!

Nah, it's fine. Just drop me at the nearest tube.

Can't tempt you back to mine for a post-shag smoke?

Ahhh.... Let's work up an *appetite* first.

What about some dinner before we...

So old-fashioned.

...and... full of... surprises....

This is hard for me to say, *I'm sorry.* You're so fucking beautiful... and...intense....

Riding up front? Must be in your good books now.

Not that you deserve to be.

What sort of vigilante surgeon would want to save the lives of criminals?

He's not *worth* your bother, if you ask me.

Ignore her, she gets like that when our supply line is down. Try this.

Nah, I only use sterilised cutlery.

Vikram, are you *okay?* You look sick?

Let me take a swab and get you checked out.

No way you're getting *my* DNA. I'm fine.

Just a cold. A very British cold!

Shouldn't you be taking care of your *bro?* His girlfiriend died, right?

Lewis's last words to me were, "Fuck off and die."

Tissue sample will do.

PING

Shit, DI Moss. And three missed calls from Lewis...

No way you'll get past airport bio-checks with that fever!

Oh, that can be arranged!

I'll go back to India tonight to shore up our supply chain, so *chill* out, Priya.

Lucky you saved our *Babba's* life or you'd be a key ingredient in our Goan fish curry!

I'd prefer death by chicken tikka masala, myself.

TB, cholera, super-MRSA!

Yet we *continue* to serve the West because we're desperate.

You *Brits* sit here in your cosy houses, with your high-tech hospitals. Yes, people are dying here--

--but in India, it's fucking carnage.

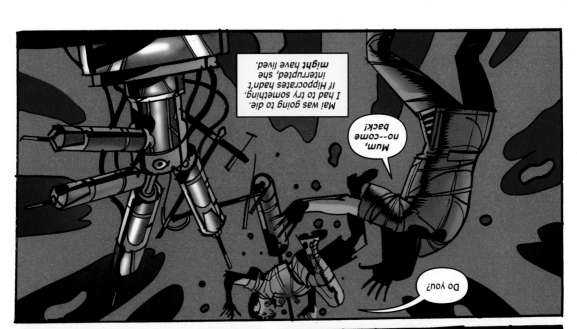

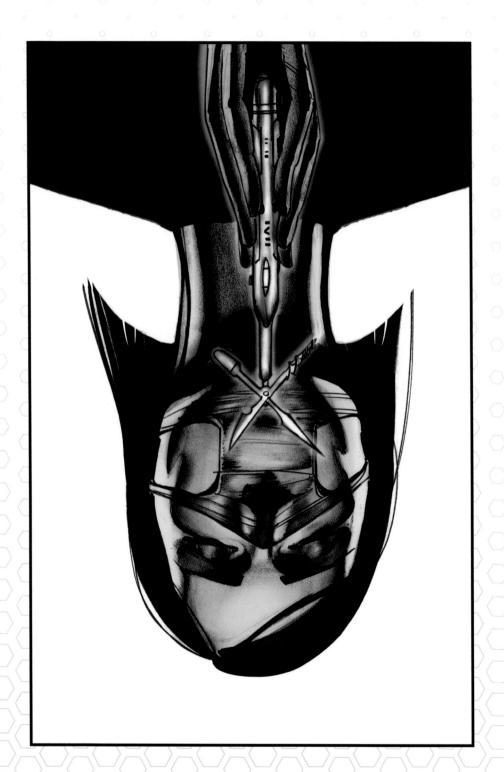

CHAPTER SIX

The hard truth is, though, as doctors, there are times when nothing is exactly what we must do.

But she would have died, right? I couldn't stand by and do nothing.

I'm 87. I want to die knowing there's *hope* for my country yet.

No one in this new government is *listening* to my concerns.

John, I'm worried for our future.

In Japan, we have much more *respect* for our elders.

I'll do whatever is needed for King and country.

Just *Charles*, please. There are very few people I can trust anymore.

I know we're considered *relics* here since independence, but I'd be quite devastated if I didn't have a home in Scotland.

What an honour to be sharing this moment with you, King Charles.

BALMORAL, SCOTLAND

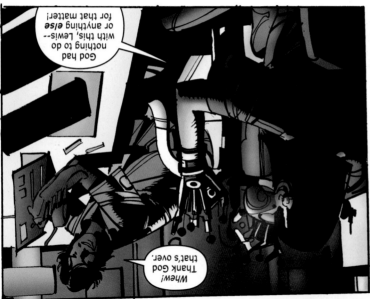

God had nothing to do with this, Lewis-- or anything else for that matter!

Whew! Thank God that's over.

Good stats. Now let's get that appendix out!

Just keep an eye on him. Everything should settle and then we can carry on.

Okay, I've pulled the port and the gas is coming out. What else?

The pressure must've stimulated the vagal nerve--let the gas out now!

He's bradycardic, not responding to fluids.

Nick, what's Chu's status?

We're back up!

Shit, we've *lost* the connection!

Nick...losing you *shash*...the pressure stimulates the vagal nerve...*shash*...and can stop the heart...

...just let the gas out...*shhhh*...*shhhhh*

What? Blood pressure is low! Can't hear you!

Lewis, *check* comms. This robot's not responding properly!

I've got the appendix. Starting to dissect the artery.

Like most things, depends on your perspective.

Yes, all good up here. Or am I down here? Can never figure *that* one out!

Looking at right iliac fossa. I see the appendix is tucked behind the caecum.

Nick, all good?

One day someone's going to say *no* to you, Rosa.

We can save them together, Sis. I know we can.

And that bird was *already* dead. Come help me, Mars.

I'm working hard here! Come in, shut the door.

Oh, my God, Rosa! What have you *done* to those poor animals?

I found them squashed under the tree. I'm trying to *fix* them. Like how Dad taught us.

Rosie! Rosie, it's lunchtime. Cheese toasties--your favourite!

I'd rather be a chicken than as *dead as a dodo!*

Mars, I love you, but you are such a chicken sometimes.

Darling Vivienne, our daughter Martha has a *wise* head on her young shoulders. Hope you realise what all this travel is doing to the girls.

John, I'm not having this argument again. Now, *back* to bed—all of you!

Not the storm, Mum. The tree that fell down and, because and, I need to....

I told her *not* to go! I didn't want her to get hurt.

Seems like I'm not the only *adventurer* in the family, John. Rosa was making a swift exit to explore the storm.

It's 2AM for Christ's sake. Your mother's flying out to Colombia tomorrow.

That's right, but I *am* the boss of you both. Get back to bed now, ladies!

I'm going outside! You're not the boss of me.

Rosa, listen to me. We swore to protect each other forever!

Let go, Martha. I want to look at the tree!

Dad told us to stay inside. It's not *safe* and you'll get us in trouble!

Noooo, Rosa! It's dangerous.

Look at that tree, Martha-- it's falling down! Let's go outside and see.

CRASH

No, it's terrifying!

What are you talking about? This is brilliant. It's epic!

I don't like it, Rosa! Draw the curtains!

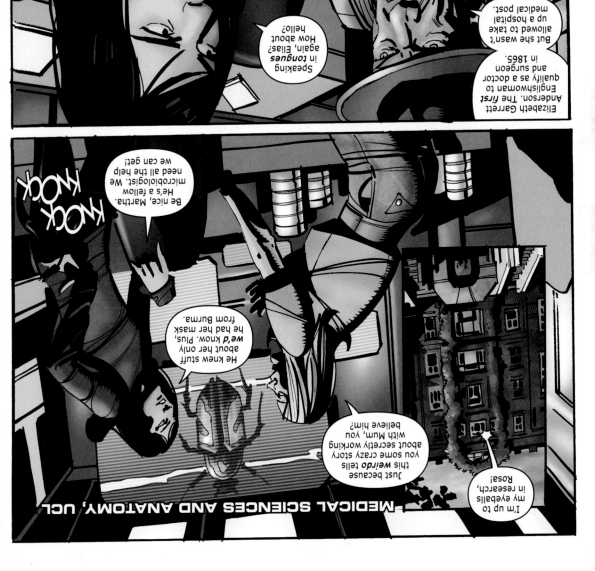

But, like my Dad says, sometimes we have to work for money, *not* love.

We've found someone. He's...how you say? Off the radar? We're negotiating terms now.

No! News of Mr. Chu's condition *cannot* be public.

We must be discreet. My team reports very few surgeons have training in *Tan Yunxian* machine.

But we're running out of time!

We've told you, robotic is our only option. Dr. Bradshaw can do set-up, but he's not a surgeon.

Beijing require we use their state VR surgical support team back on Earth.

4D ultrasound scan shows acute appendicitis, Mr. Ho. It'll *rupture* if we don't operate soon.

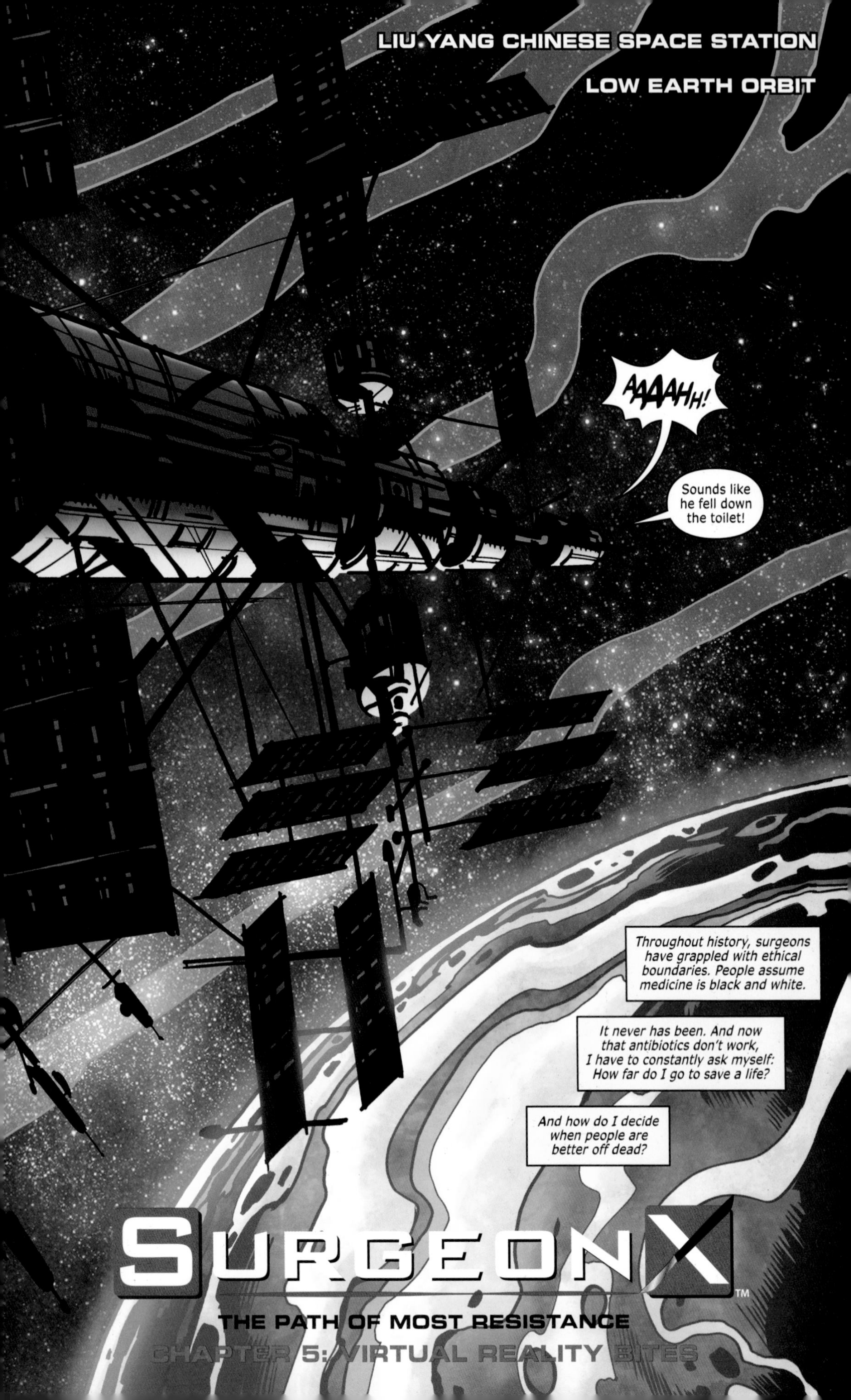

THE PATH OF MOST RESISTANCE

SURGEON X ™

Throughout history, surgeons have grappled with ethical boundaries. People assume medicine is black and white.

It never has been. And now that antibiotics don't work, I have to constantly ask myself: How far do I go to save a life?

And how do I decide when people are better off dead?

Sounds like he fell down the toilet!

AAAHH!

LIU YANG CHINESE SPACE STATION

LOW EARTH ORBIT

CHAPTER FIVE

Did you say *Hippocrates?*

It's called *Commentary on the Aphorisms of Hippocrates.*

Wait, could he have *known* Mum? He's hidden some of her old encrypted computer files inside this digital book.

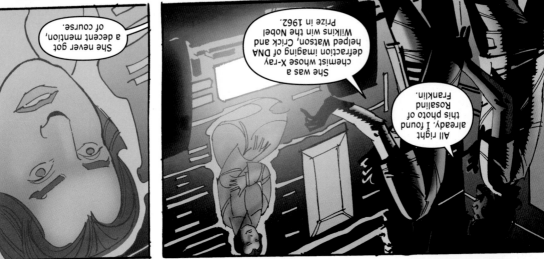

She never got a decent mention, of course.

She was a chemist whose X-ray diffraction imaging of DNA helped Watson, Crick and Wilkins win the Nobel Prize in 1962.

All right already. I found this photo of Rosalind Franklin.

"My sister can have the pleasure of figuring you out!"

C'mon, Lewis!

Chill, Rosa. Been deciphering the contents of your mysterious package for hours. Need some fuel here.

Bugger that! Think I had that luxury when scraping *rotting flesh* from your friend's leg?

I know. I'm sorry. We're doing our best.

What's wrong with them stupid, stupid, stupid people? HURRY UP!

The scientists need to hurry up! What's *taking* them so long?

My baby sister died two years ago. Sepsis. Mummy *still* cries.

My Gran *died* last year when she was on chemo. Got an infection and the antibiotics didn't work.

Some people are working on that--the Russians are making ground with *bacteriophages*, viruses that kill bacteria.

Why doesn't someone make a *super* antibiotic, which doesn't stop working? Ever.

Shit, it's DI Caspar Moss. Why'd you mess with a *murder* scene, sis?

And that film of Mum--I can't get it out of my head! Do I have to watch it again?

I need you to look for *clues,* Lewis. You're good at that stuff.

Now, put the kettle on. Detective Moss has always had a *thing* for me. So, I'll do the talking.

You seem *better* lately, but I constantly worry. Still hearing voices?

Yes, sometimes, but it's okay, sis. I've got *them* under control.

BRIING BRIING

That's good, now get the door! I gotta put on some makeup to disguise my barefaced lies.

Don't know, but beetles almost never go extinct. So maybe they can *teach* us a thing or two.

This is too much, Rosa, we need help with this! I've got children and--

Omigod-- what the fuck?! *More* beetles? Grab them.

But why is it on us? What are we getting ourselves into?

Let's just *take* this shit and get the hell out of here so maybe we can try to figure this out!

Me, too. But, Martha, only you, Lewis and I can know about this. We can't go to the police.

Oh, Rosa. She looked so real. God, I feel sick.

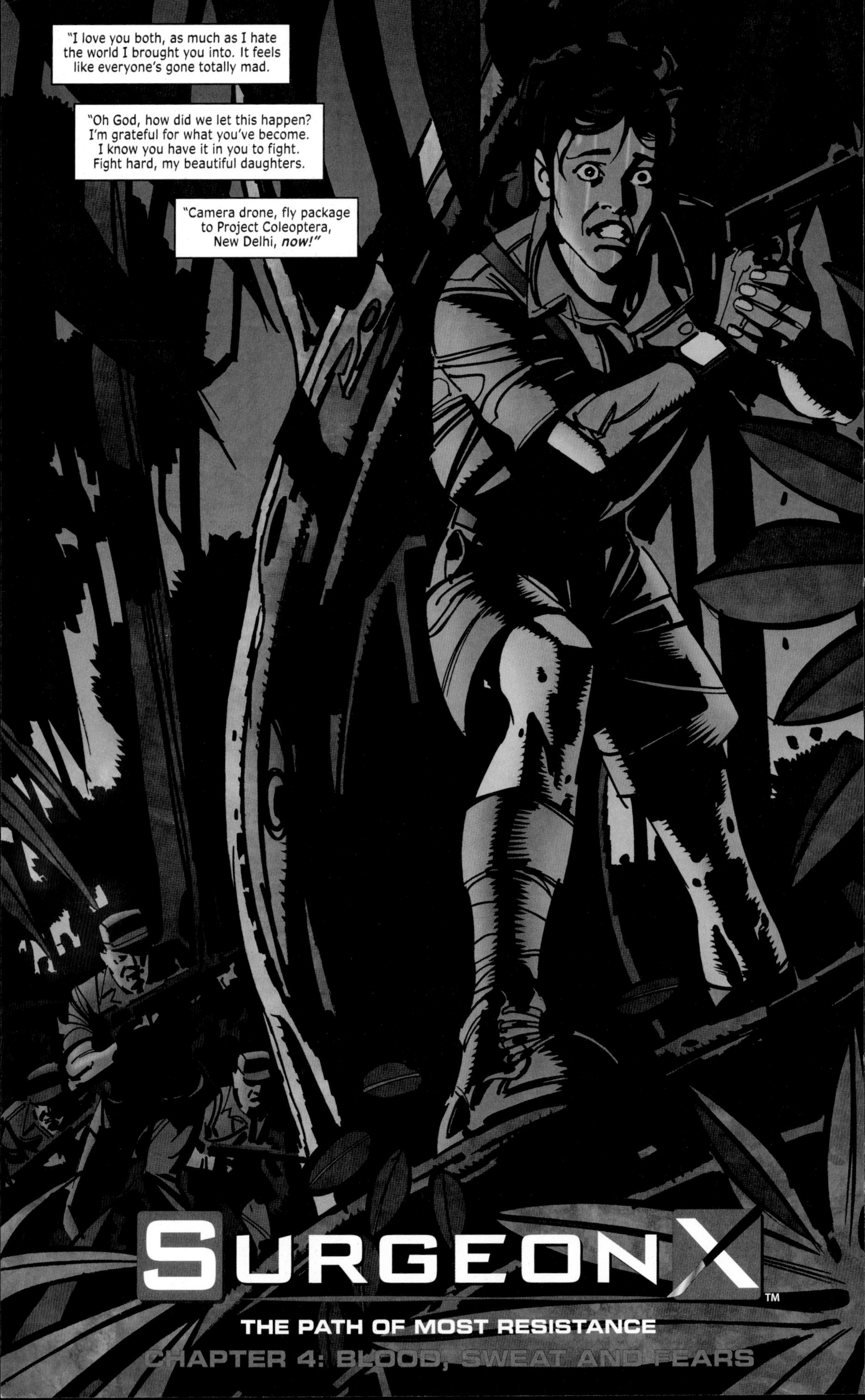

SURGEON X ™

"Camera drone, fly package
to Project Coleoptera,
New Delhi, now!"

Fight hard, my beautiful daughters.
I know you have it in you to fight.
I'm grateful for what you've become.
"Oh God, how did we let this happen?

like everyone's gone totally mad.
the world I brought you into. It feels
"I love you both, as much as I hate

Mum?!

A Holorug. Let's roll this out and see what Dana was up to.

I didn't mean to another part of the building!

Here we go.

Dana left a trail, and we owe it to her and Mum to follow it.

Look at this.

Invertebrate Sample Lab Row 64 Drawer 19

Rosa! Dana's been *murdered!* Please, let's move.

This is *wrong!* We should be calling the police, Rosa!

Looks like she left a message.

Martha, get the lights. Remember Dana's chemistry lesson on invisible inks?

Her office is on the top floor. Now, let's *move* before *Bionic Bertha* stops us.

Hello, Dana... are you okay?

Dana...?

Omigod, Rosa. Is she dead...?

Yes, *fuck!* Quick. Close the door.

Are you nuts? Let's call the police, *NOW!*

Hold on a sec. Dana called us here because she knew something about Mum.

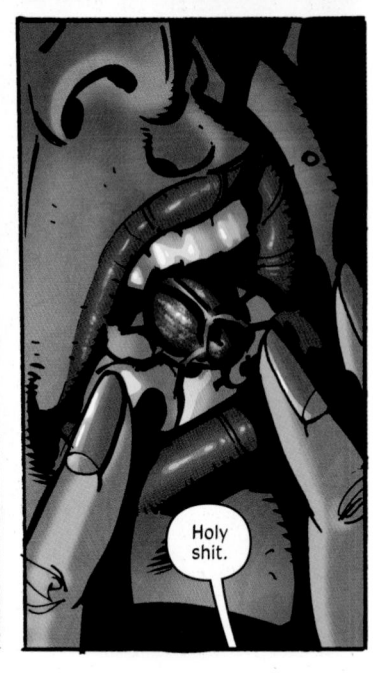

Holy shit.

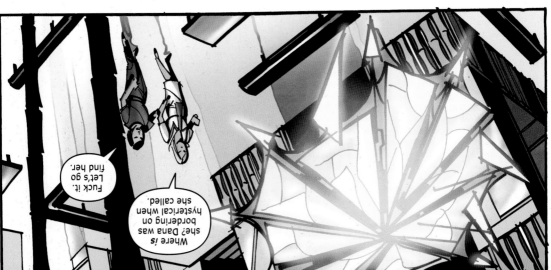

Can you believe it? Some freaky *flesh-eating* bacteria. Twenty hospitalised so far.

Blimey. Heard they traced it back to a hot tub party in Camberwell.

Wife wants a hot tub. Maybe I'll *get* her one after all!

I told you, Lewis. My *PCI* score's too low-- they'll shove me in a corner and let me die!

What about your sister?

I always wanted to be like my Mum. At 18, she came to Imperial College on a science scholarship, later winning the Fleming Prize for Excellence.

By 2025, she was a revered professor, travelling Northern China treating thousands dying from a resistant-TB outbreak.

You were drunk, too bad. Loose lips, man. This thing's *eating* me alive!

Saint Vivienne, Martha and I used to call her. There's no way I could live up to that.

Fucking better get your bloody criminal sister to *help* me, or else her secret is out!

So, I came up with something different.

NUELLA POWELL WINS LONDON MAYORAL RACE

IN HISTORIC ELECTION WIFE OF MURDERED CANDIDATE OVERWHELMS OPPOSITION

The Lionheart Party candidate Nuella Powell was swept into office by the voting power of the right-wing city establishment. But in a shocking outcome, she received 35% of minority votes, specifically those originally hailing from Northern Africa, Syria and climate refugees from Bangladesh and Pakistan. Experts believe this unprecedented voter shift was due to Mrs Powell's defense of migrant groups against widespread bomb-related accusations.

Prime Minister Nicholas Goodwin congratulated Mrs Powell on Chatterblam with the campaign's hashtag #therightchoice.

Mrs Powell, dubbed 'The Iron Widow,' won the Lionheart Party mayoral nomination after the death of her husband, Jim Powell, in the tragic City Hall bombing five months ago.

Mrs Powell, always sharp suited, is a former stock trader who campaigned heavily in support of the Antibiotic Preservation Act and has pledged to uphold the rationing measures in spite of public unrest and violent demonstrations.

EXPLOSIVE REVELATION
LONE CABBIE RESPONSIBLE FOR BOMBINGS

Claims Revenge Against Taxis with Auto-Drivers

For the past year London has been terrorised by devastating bomb attacks. Two days ago Detective Inspector Caspar Moss revealed that his team had caught a suspect and were questioning him in relation to a total of five bombings. Moss confirmed late last night that the bomber is John Smith (51) from Homerton, a former black cabbie who was protesting against driverless taxis.

It is rumoured that Detective Inspector Moss was nearly pulled off the case for claiming the bomber was homegrown and not an extremist. A decorated soldier who fought in Europe and the Middle East, Moss is now being recommended for the Commissioner's 'Total Excellence in Policing' award.

MYSTERIOUS DOCTOR DUBBED 'SURGEON X'
Dangerous Vigilante or Saviour?

Reports of the renegade surgeon whose identity remains unknown have escalated as increasing numbers of the population are disqualified from antibiotics because of low PCI scores. Nothing is known about the mysterious surgeon whose clandestine surgery has eluded officials.

One patient, quoted anonymously, speaks of dark web contacts and procedures conducted in a secret location, but has only praise for the medical services.

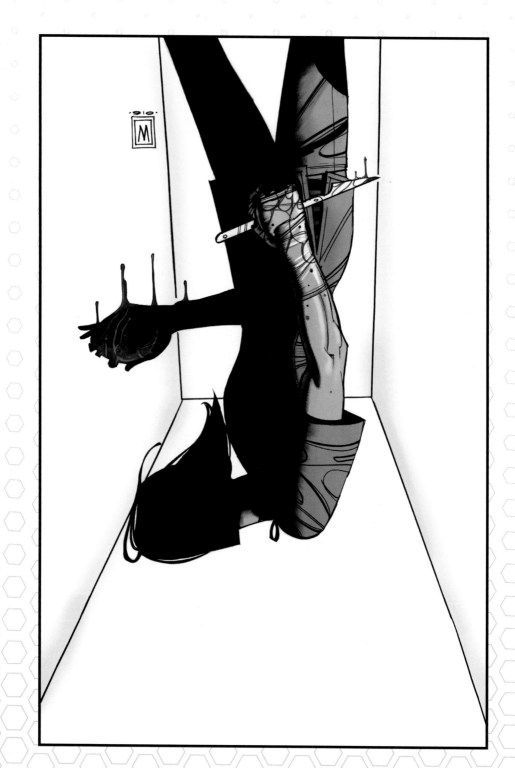

CHAPTER FOUR

Hate to say this, Caspar, but I think I need your help.

Might be time to pay a visit to my surgeon's graveyard again.

What the hell?

Why is your *head* attached to that monkey's body?

Omigod.

Want some help?

I will keep pure and holy both my life and my art....

No. Get *out* of my car, detective!

Get over yourself, I'm following up on a lead. Why are *you* at Madison Smith's house?

Well, I, uh...

WEEOoo WEEOoo

Detective Moss, why are you tailing me?

My car alarm's nuts. Miss those good old cogs and wheels.

WEEOoo WEEOoo

You nearly fucking killed those people!

Hippocrates, show your face, coward.

Could this day get any bloody worse...?

Command override!

What the hell?! Divert!

When your autonomous car is about to hit a crowd, should it swerve and kill the driver instead?

Bollocks, get off my tail!

It still feels surreal sometimes. But life goes on....well, some lives go on anyway.

I look back at my younger self and in my wildest nightmares, I didn't see this world coming.

What was that?

KRASH

Touché. Let's take a look, age before beauty...

Said the paranoid schizophrenic...

Came from downstairs. You've become *massively* neurotic since you had kids.

No sign of her at all. Really thinking we should call the police.

I keep telling you, we go to the police, the Varmas will kill her. I actually *prefer* the uncertainty right now.

Rosa's been gone since yesterday. That's *weird*, right--nearly 24 hours?

Let's go inside. Got my spare key.

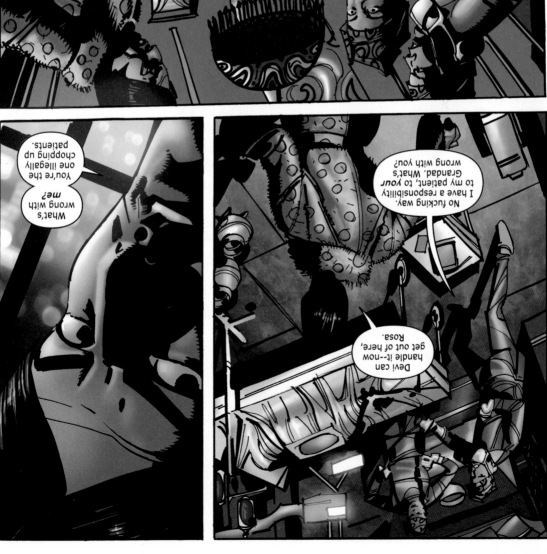

He nearly *died* under your care--he's sure as hell not going to die under mine.

Hold on, I need to monitor him for 24 hours.

Car is waiting. You'll have to wear the bag over your head again. Can't risk Babba's security.

No, it's *time* for you to leave! All the drugs you asked for are in the flight bag by the front door.

Since you won't let me take your Babba to a hospital, I'll stay and observe him here.

The ketamine will be wearing off soon.

Chest drain is bubbling and swinging nicely, obs are all good. Let's run a blood gas and check that.

Lewis, calm *down*, I'm sure Rosa's fine. What was it that triggered these thoughts? Talk me through it.

I can't tell the police or it will make it *worse!*

She's a *prisoner*, held at gunpoint!

Rosa...she's a prisoner...this Indian restaurant... you see...and they took her!

Lewis, what are you *doing* here?

Excuse me, class.

Students, I know you'll be devastated, but tonight's lecture is over. See you next week.

Martha, *sorry* to interrupt!

Believe it or not, these are some of the places scientists have looked to find new antibiotics.

--alligator blood, underground caves, marijuana, panda DNA, *even* old TVs.

Cockroach brains--

CRACK

I should know.

"A place of bitterness and regret, where he must look for an explanation for his failures." *Rene Leriche,* second-year required reading.

People put surgeons on a pedestal, because the alternative is too frightening. Truth is we're human and we make mistakes.

He'll be *fine* with one lung, Devi.

"Every surgeon carries within himself a small cemetery, where from time to time he goes to pray.

I've had to re-train in many of the old surgical techniques.

So Bebba will only have *one* fully working lung now? It's all my fault!

He'll be more useful unconscious. These procedures are brutal. Mouthy men are *always* first to faint.

Rosa Scott, we meet again. You two *are* related....NHS employee records.

Insubordination, inappropriate use of resources, rudeness. We have a lot in common.

Resigned, **28 Sept. 2036.** Last patient-- Madison Smith.

Just what are you up to *now*, Miss Scott?

Okay, Lewis Scott, got some *priors* here--disturbance of the peace, trespassing.

Huh? *What's this* file: no access bullshit?

That's right, *run* away, you fucking runt!

I'm on your side, Constable, but--

You attack a *vulnerable,* sick man in front of me again and I'll pound your head into that pavement, got it?

Y-yes, sir.

IP over avian carriers. Five terabytes of data is faster and safer via pigeon than London's creaking internet!

There you are, just as scheduled.

cOOOo
cOOO
cOOo
cOOOo
cOOO

Row two, seat nine.

Good, everything's here.

Gotta get the old networks going again.... need them.

Mum's gone, now Rosa's gone.

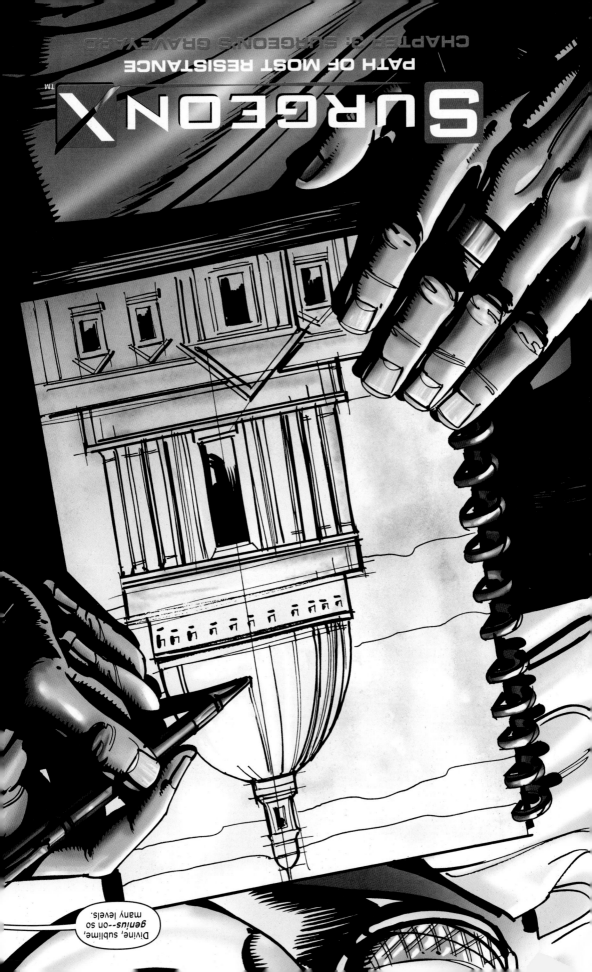

Poor doctor, does it *hurt* when I press *here?*

"Mmfff!..." Can't...

Shut up! You surgeons are total control freaks.

What the hell is *that?*

Good. You'll be needing it.

My Diagnostics *Dragon.* A doctor's lifeblood.

That's a laugh. Your phone is older than my Grandad. Got any other tech?

Vikram, you *can* trust me. We can make some serious money.

I'll get help, Rosa!

Lewis, it's okay, run!

Get out of here, before I *deep-fry* your arse!

What's going on? Take me, instead!

Vikram, *take her!*

And careful with the *hands*-- she'll need them.

CHAPTER THREE

I wish everyone would just shut the **fuck** **up**, so I can hear my thoughts clearly!

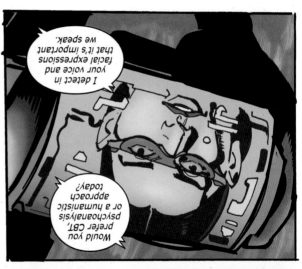

I detect in your voice and facial expressions that it's important we speak.

Would you prefer CBT, psychoanalysis or a humanistic approach today?

You're **not** real. Leave me alone!

Lewis, it's Emma here, your **virtual** therapist. Your sister said you might need someone to talk to.

ping

Those fascists killed you. I won't let them hurt Rosa or Martha, I promise.

I'm sorry I couldn't help you, Mum.

I tried to find you. I wrote to you every day. Here's another letter.

FOR MUM

BUCKINGHAM PALACE, WESTMINSTER

The Great British honours system, where the worthy *and* the super-rich are rewarded.

So, Dad, should we now refer to you as "Commander of the British Empire"?

"Sir" sounds better. I *should've* been knighted. Perhaps next time?

Now, get to your seats. I need to find your brother.

Can't believe Charles still spouts on about that *homeopathy* bullshit. His pissy arts degree doesn't make him an expert on anything.

There you go again. But not as eloquent as your post on that anti-vaxxer's blog-- *"fuck off and learn how science works."*

I *stand* by that line, Martha.

That's my sister. Winning hearts and minds, one *insult* at a time.

I don't advocate homeopathy, but not everything on the fringe is bullshit. And, I'm the *microbiologist* in the family.

Try opening your mind to *new* ways of thinking and even *you,* Rosa Scott, might learn something!

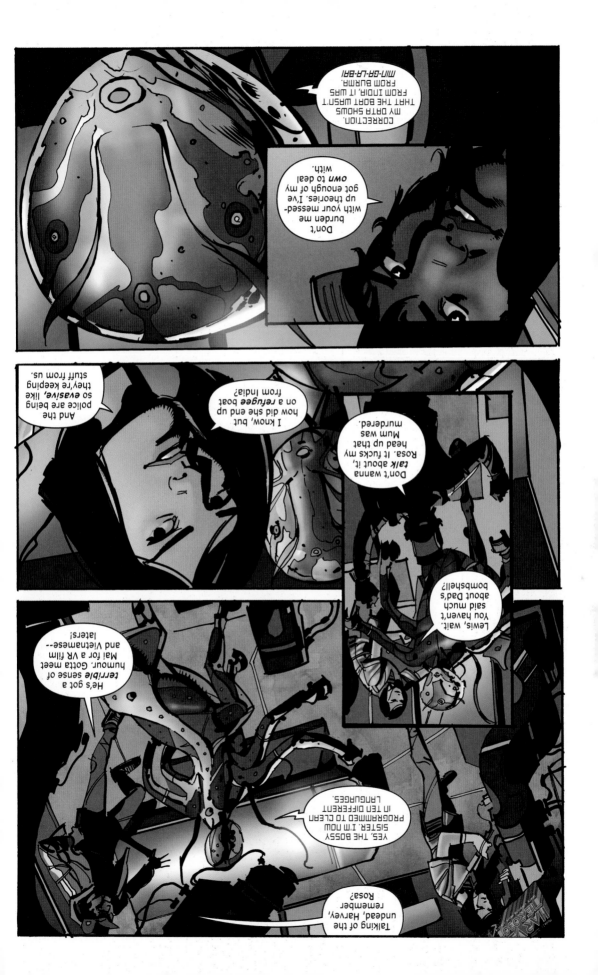

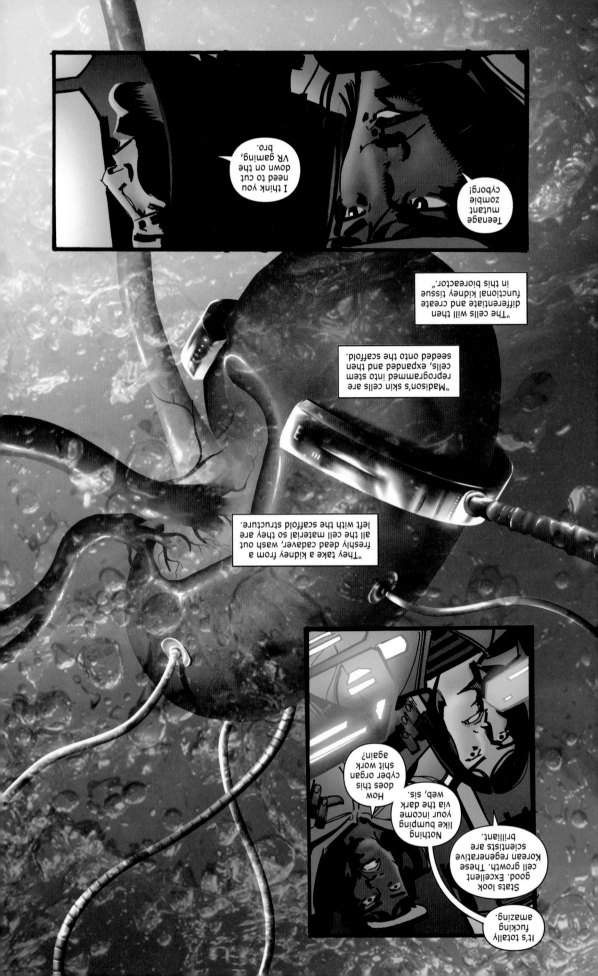

Not all of you are drug dealers, rapists or disease-infested--some of you are *actually* good people.

Remember-- "Vote Nuella, the *right* choice, the people's voice!"

I am pleased to announce that the Metropolitan Police have a reliable lead, which suggests you *maligned* minorities might not be the culprits.

God, I'm getting good at this!

It takes a real pro to sound sincere with the smell of greasy samosas wafting up one's nose!

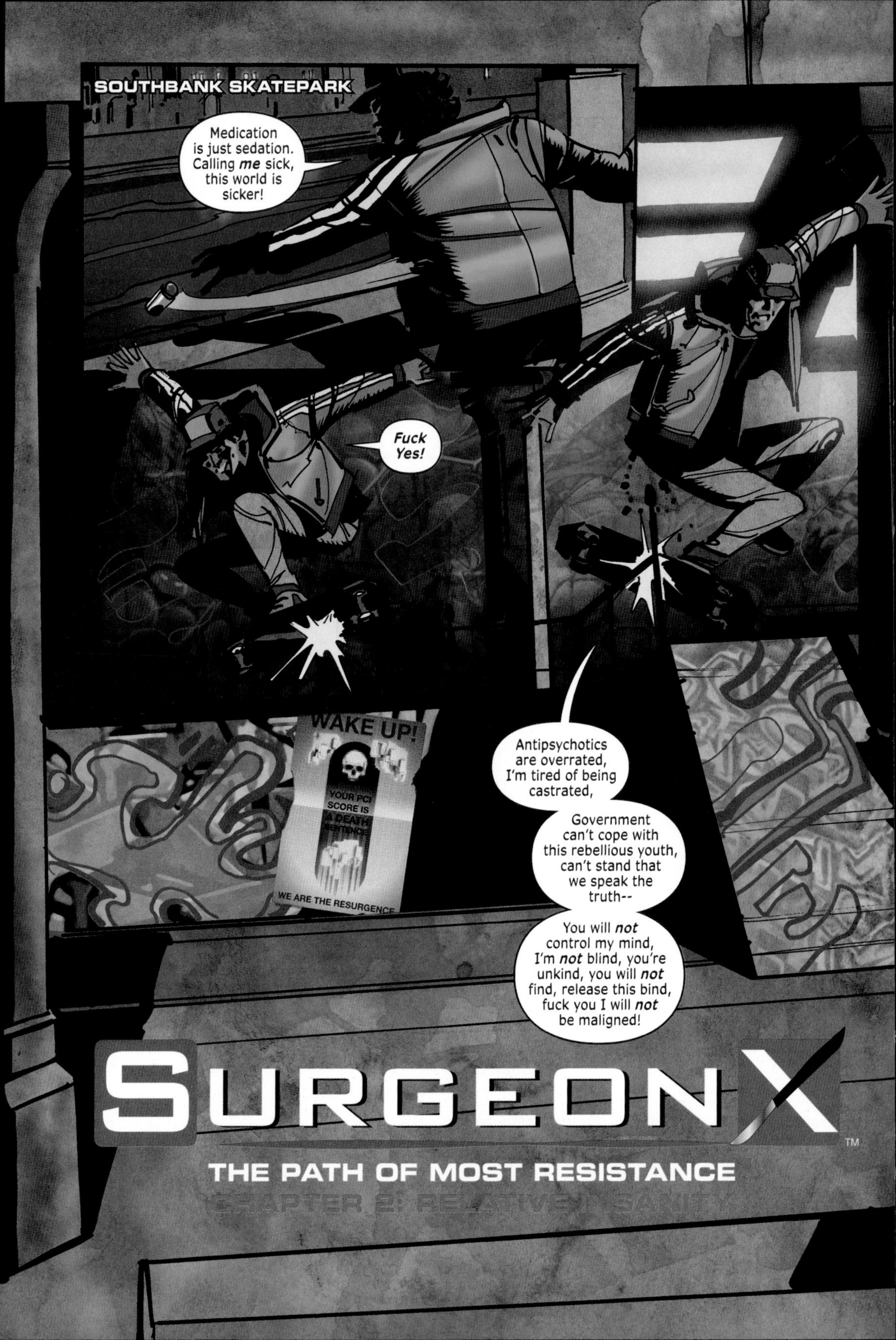

THE PATH OF MOST RESISTANCE

SURGEON X ™

Antipsychotics are overrated, I'm tired of being castrated.

Government can't cope with this rebellious youth, can't stand that we speak the truth--

You will *not* control my mind, I'm *not* blind, you're unkind, you will *not* find, release this bind, fuck you I will *not* be maligned!

WE ARE THE RESURGENCE

YOUR PCI SCORE IS A DEATH SENTENCE

WAKE UP!

Fuck Yes!

Medication is just sedation. Calling *me* sick, this world is sicker!

SOUTHBANK SKATEPARK

OR RISK LOSING YOUR LIMBS

WASH YOUR HANDS

Some people won't like
it--I don't give a shit.
Extreme circumstances
call for extreme medicine.

So no more obsessing
over futile things--over
not having control--over
not treating patients in
a way I truly believe in.

I'm more afraid of
living--of wasting
what time I have left.

I've seen so much
death I'm actually no
longer afraid of dying.

APOLLO
MEDICAL

What really
happened. Your
mother deserves
to have the truth
known about
her death.

She was
injected with a
paralysing drug,
Succinylcholine.

She couldn't
swim to safety...
she was
murdered.

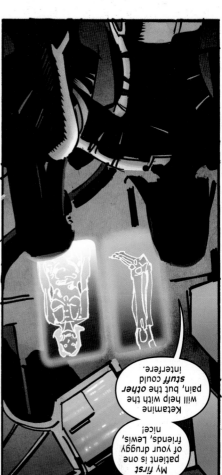

"For decades, big pharma failed to *invest* in new antibiotics--not profitable. And governments didn't devote enough resources until it was too late."

"Hospitals have become *breeding* grounds for these deadly infections, and enforced home isolation is now the *only* option for the seriously ill."

"It changed our world."

"But, it wasn't until 1942 that penicillin became available for use, saving several hundred thousand lives during World War II."

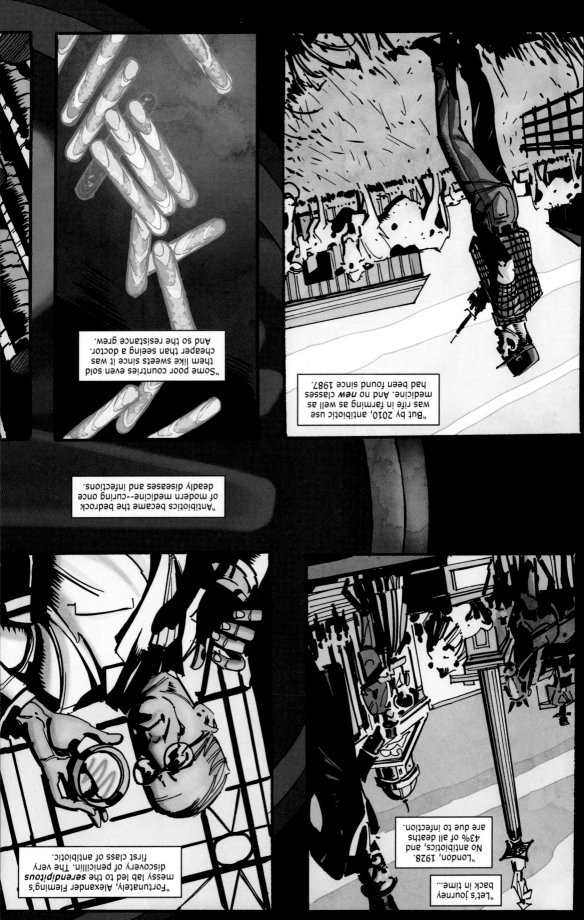

"Some poor countries even sold them like sweets since it was cheaper than seeing a doctor. And so the resistance grew.

"But by 2010, antibiotic use was rife in farming as well as medicine. And no new classes had been found since 1987.

"Antibiotics became the bedrock of modern medicine--curing once deadly diseases and infections.

"Fortunately, Alexander Fleming's messy lab led to the *serendipitous* discovery of penicillin. The very first class of antibiotic.

"London, 1928. No antibiotics, and 43% of all deaths are due to infection.

"Let's journey back in time...."

Poor girl. She doesn't stand a chance in this fucking mess of a world!

WASH YOUR HANDS

OR RISK LOSING YOUR LIMBS

Today is my last day here, but perhaps I can still help you.

I'll be in touch soon.

Don't make excuses! Your daughter's *life* is at stake.

Processed food is cheaper and I read that it's safer.

Her PCI score has to come down. You need to cut out the junk food.

And that her weight isn't helping.

The trainee said she could get an infection from the transplant.

Mr. and Mrs. Smith, we need to talk.

She's a Leo, like her Dad. Thank you for saving her.

Your daughter is a brave and strong young lady.

You *should* have waited for me, Rosa.

You're a class act, Hugo Talbot. Calling in management.

Professor Talbot might've saved that kidney. She'll be waiting years for a transplant.

We can give Madison a home dialysis machine until we get her a new kidney.

I was on my way. You should have waited.

She'd be *dead* if I waited. She was bleeding out.

She's high risk for an infection, disabled and obese. She may not qualify for more antibiotics.

You saying I should have let her die?

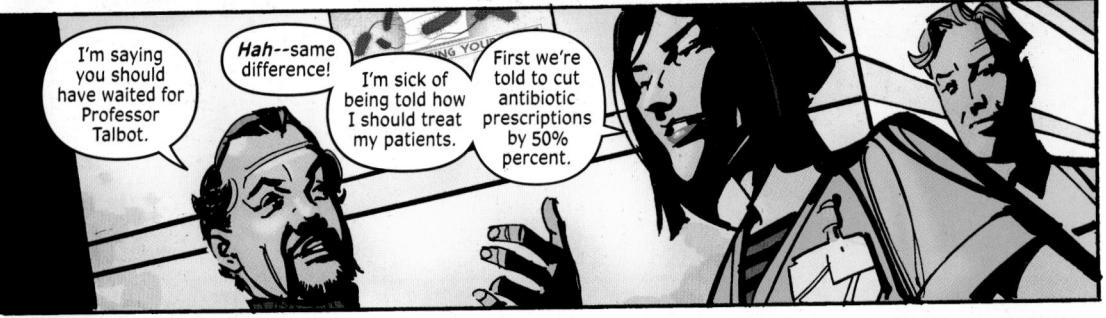

I'm saying you should have waited for Professor Talbot.

Hah--same difference! I'm sick of being told how I should treat my patients.

First we're told to cut antibiotic prescriptions by 50% percent.

And next the government locks up the antibiotics and *controls* their use?

These *shifting* rules are never-ending. Enough!

Doctors should be in charge of anti-biotics, not the *"Preservation Bureau."*

Fuck this shit.

Rosa! You can't talk to me like *that*.

I can if I quit! You can shove *that* up your bureaucratic arse!

I heard about the blast at *St. Pancras* station!

It's exciting, isn't it? All that *blood* and *broken* bones.

Yes, a *real* party. Is that patient prepped?

Nice to see you *too*, Miss Scott. She's the first in, and only has one kidney.

Heard about your Mum, sorry.

Well... thanks.

Now, let's try to keep *this* one alive. Status?

Her obs *were* fine, but now her pulse is up and blood pressure down.

Signs of internal bleeding. Prep for surgery *now*.

New government guidelines on antibiotic restrictions are in.

Not interested. My first priority is my patient.

If someone needs antibiotics, I'll prescribe them.

Pointless. They're under lock and key. We gotta follow protocol.

Since *when* did doctors take orders from politicians? Fuck that!

SURGEON X ™

THE PATH OF MOST RESISTANCE

As a doctor, this thought scares the shit out of me.

I now believe life is a privilege, not a right.

LONDON, 2036

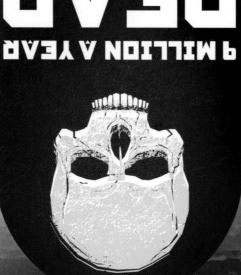

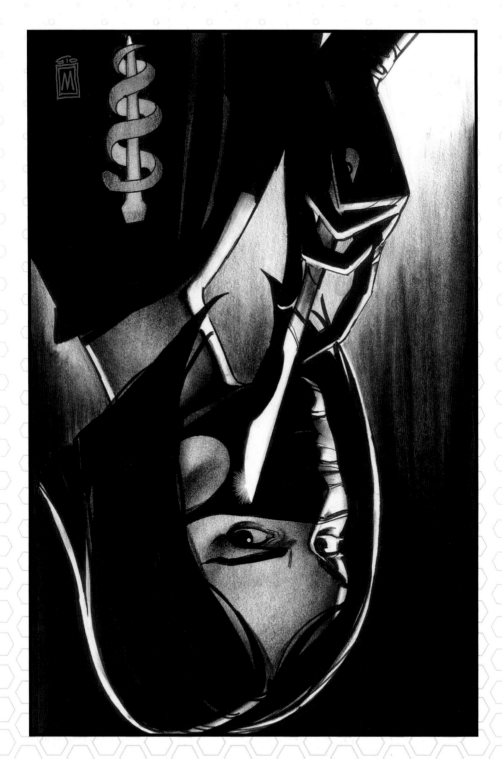

CHAPTER ONE

In Memory of
John Watkiss

In appreciation of
the staff at the NHS and
experts working in the fight
against antibiotic resistance

Logo/Book Design
RICHARD BRUNING

Wellcome Trust is an independent
global charitable foundation
dedicated to improving health.
They support bright minds in science,
the humanities and the social sciences,
as well as education, public engagement
and the application of research to medicine.

Key Consultants
Dr Matthew Avison (Microbiologist)
Dr Vaughan Bell (Neuroscientist & Psychiatrist)
Prof Laura Bowater (Microbiologist)
Dr Elvira Bramon (Neuroscientist & Psychiatrist)
Ms Laura Coates (Surgeon)
Mrs Marylyn Cropley (Social Worker)
Prof Sarah Cunningham-Burley (Sociologist)
Dr Lindsay Evans (Chemical Scientist)
Dr Helen Fletcher (Immunologist)
Dr Jess Healy (Pharmaceutical and Biological Chemist)
Dr Rob Hughes (Diagnostics Engineer)
Dr Gavin Jell (Tissue Engineer)
Prof Alan Johnson (Clinical Scientist)
Prof Roger Kneebone (Surgical Education & Engagement Science)
Prof Helen McShane (Immunologist)
Ms Caroline Moore (Surgeon)
Dr Catherine Mohr (Surgical Roboticist)
Prof Adrian Mulholland (Chemical Scientist)
Mr Nick Newton (Surgeon Lieutenant Commander)
Dr Harriet Palfreyman (Historian)
Dr Emmanuelle Peters (Psychologist)
Dr Lisa Pierre (Pharmacologist)
Dr Adam Roberts (Microbiologist)
Dr Emma Sutton (Historian)
Paul Walker (Chemical Scientist)
Dr James Wilson (Philosopher & Ethicist)

Special Mention
Simon Armstrong (Poster Design, Animation)
Marnie Chesterton (Audio Documentaries)
Dr Duncan Copp (Producer)
Oliver Kenney (Composer, Playlists)
Michelle Martin (Audio Documentaries)
Michael Mensah-Bonsu (App Designer)
Rob Pierre (Jellyfish Digital)
Kamini Plaha (Researcher)
Thy Quach (Film Editor)
Prof Alice Roberts (Narrator, Anatomist)
Silverglade (Post Production)
Dana Stevens (Social Media)

Wellcome Trust
Dr Tom Ziessen
Dr Tom Anthony
Angela Saward

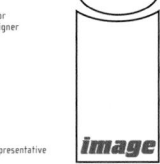

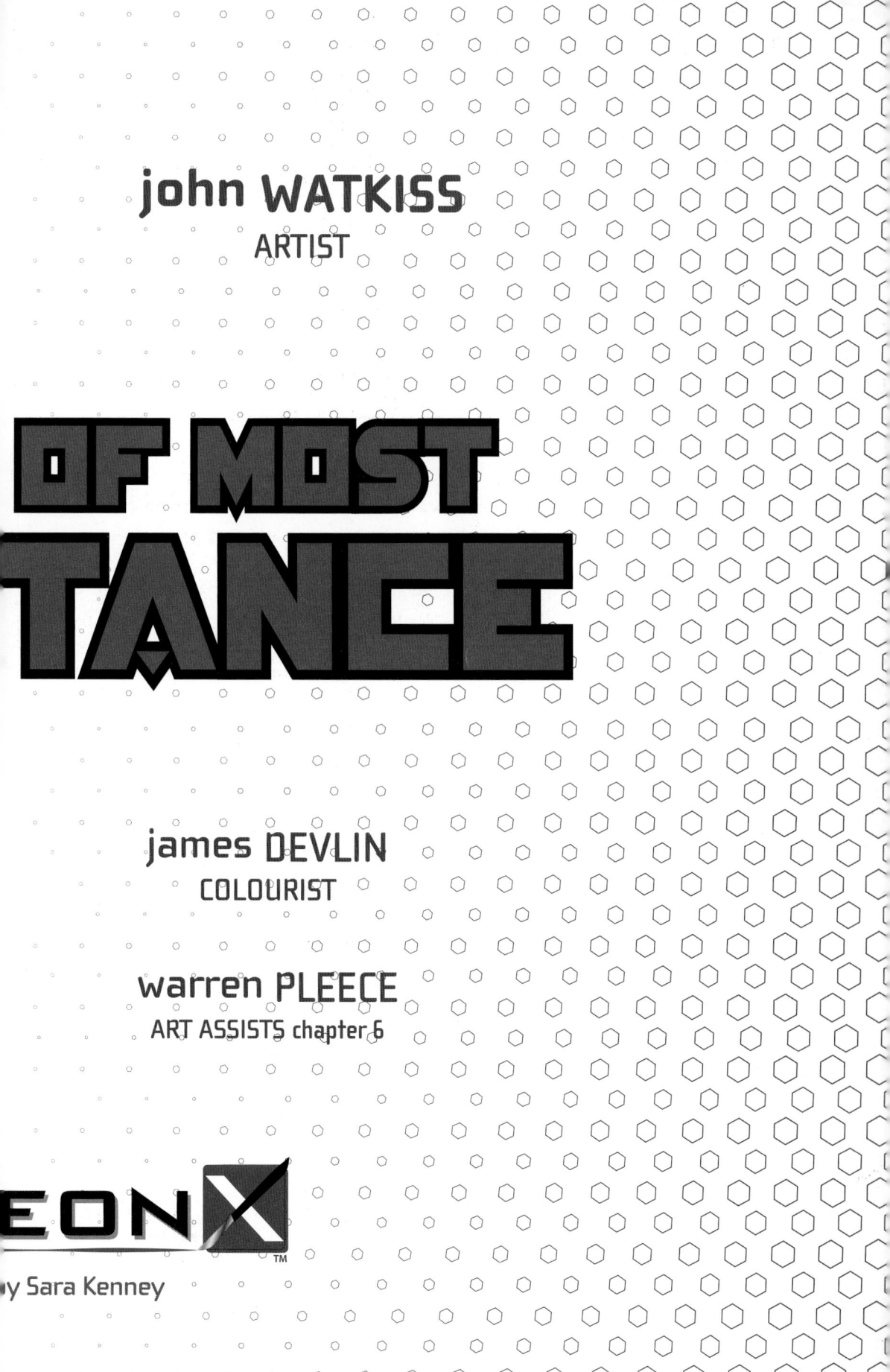

SURG
Surgeon X create...

jared k FLETCHER
LETTERER

karen BERGER
EDITOR

THE PATH
RESIS

sara KENNEY
WRITER

SURGEON X ™

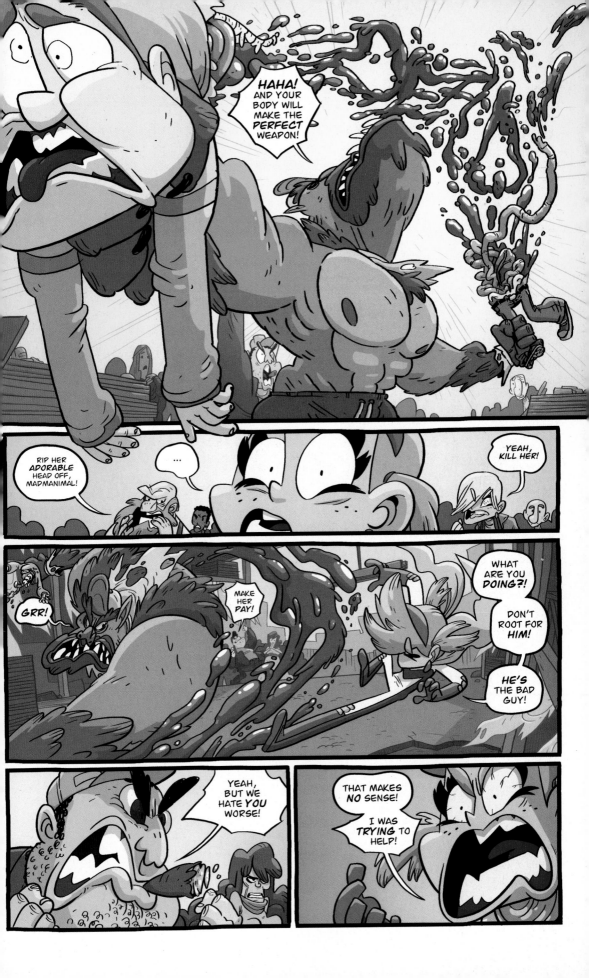

NO-NO-
NO-NO-NO!
STOP!

HULKTRESS
IS **DEAD**.

**YOU'RE
FREE.**

I'M **ONLY**
HERE TO **PROTECT**
THESE PEOPLE...

JUST
GO!

TEMPTING...

...BUT
I'D MUCH
RATHER
KILL YOU
ALL BEFORE
I GO!

**THAT'S
ENOUGH!**

HWEH!

AM I OKAY?!

SLAP

YOU JUST **KILLED** MECHANIX!

...AND THE **HULKTRESS**...

LIKE A **REAL** JERK!

OW...

AND I **ALSO** SAVED YOUR **LIFE!**

YEAH... BUT **THEY** WERE MY **FAVORITES**...

OKAY, FINE! I'M SORRY!

I'M **SORRY** I SAVED **ALL OF** YOUR **LIVES!**

YOU SUCK!

HOW IS IT THAT YOU DON'T **UNDERSTAND?!** I'M THE **GOOD GUY!**

YOU SHOULD **KILL** YOURSELF!

STOP!

WHO THE--?

ISSUE ONE VARIANT COVER BY RYAN OTTLEY & NATHAN FAIRBAIRN

CHAPTER TWO

BAY CITY, THE CAPITAL CITY OF *LAVAPLOP ISLAND*, BURNS.

AND WE HAVE OUR HERO, *GAMMA RAE*, TO THANK FOR THAT (IN CASE YOU MISSED OUR LAST ISSUE).

AND IF YOU'RE THE KIND OF *SICK ANIMAL* THAT WOULD ACTUALLY THANK GAMMA RAE FOR THAT, THE OVERGROWN REMAINS OF *BOLLINGER PARK* MAY BE THE PERFECT PLACE TO VIEW THE DESTRUCTION.

BUT NO ONE REALLY COMES TO BOLLINGER PARK ANYMORE.

WHICH IS JUST ONE OF MANY REASONS IT SERVES AS THE PERFECT LOCATION FOR A SECRET ENTRANCE TO A HIDDEN FORTRESS.

BUILT LONG AGO BY LOVING PARENTS TO PROTECT THEIR CHILDREN FROM THOSE WHO WOULD PERSECUTE THEM.

BEFORE LEAVING THEM ON THIS WORLD ALONE.

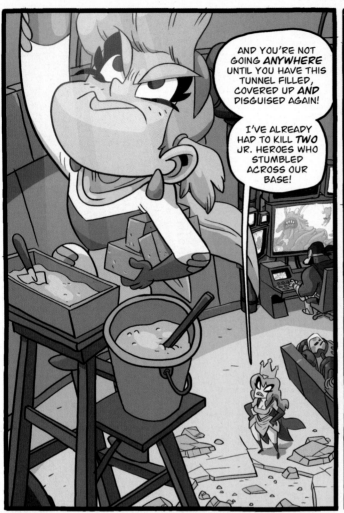

AND YOU'RE NOT GOING **ANYWHERE** UNTIL YOU HAVE THIS TUNNEL FILLED, COVERED UP **AND** DISGUISED AGAIN!

I'VE ALREADY HAD TO KILL **TWO** JR. HEROES WHO STUMBLED ACROSS OUR BASE!

DON'T WORRY, NO ONE'S GOING TO FIND YOUR **STUPID** BASE AGAIN.

I'VE BEEN THINKING...

MAYBE I SHOULD JUST MOVE BACK IN WITH YOU GUYS...

GIVING UP ON YOUR PATHETIC DREAMS OF BECOMING A HERO ALREADY?

DIDN'T SEE **THAT** COMING.

SHUT UP, **SLUDGE!**

WHAT'S WRONG, RAE?

ALL WEEK, I'VE DONE NOTHING BUT TRY TO SAVE PEOPLE AND FIGHT CRIME...

EVEN TODAY, I TOOK DOWN A **GIANT MONSTER** WHEN NO ONE ELSE COULD... AND DO I GET THANKED? **NO!**

...

OH, YOU MEAN **THIS?**

--IN THE MIDST OF THE BRAWL, A NUMBER OF HIGH-PROFILE SUPER-HEROES FELL VICTIM TO HER **CARELESS** APPROACH AND THE CITY IS NOW IN FLAMES.

WHOEVER THIS NEW **"ZERO"** IS, ALL WE CAN SAY FOR SURE IS THAT SHE REALLY LET EVERYONE DOWN...

...HER PARENTS SHOULD BE ASHAMED.

OH, SHIT...

HAHA! THAT WAS **YOU?!**

THESE HEROES THINK FOLLOWING THE RULES IS MORE IMPORTANT THAN SAVING **ACTUAL** PEOPLE!

I WAS **TRYING** TO HELP!

YOU THINK I **WANTED** THOSE WORTHLESS LOSERS TO DIE?!

NO!

IT'S NOT LIKE **I** GET ANYTHING OUT OF THIS CRAPPY CITY BEING DESTROYED!

HELLO...?

WHAT THE--?!

HELP ME... YOU'RE HERE TO HELP ME?

I THINK MY BODY IS EATING MY INSIDES...

CAVERN KID?!

YOU FOUND THE ENEMY LAIR AFTER ALL!

OH, IT'S *YOU*...

YEAH... I FOUND IT ALRIGHT.

AND NOW YOUR CRAZY BROTHER IS DOING EXPERIMENTS ON ME.

WHY DIDN'T YOU *TELL ME* THIS WAS YOUR FAMILY'S BASE?

WHY WOULD I EVEN--

WAITAMINUTE!

SLUDGE! YOU HAVE TO BE MORE CAREFUL! IF THE OTHER HEROES FIND OUT I'M RELATED TO YOU GUYS--

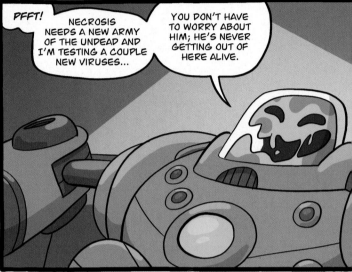

PFFT! NECROSIS NEEDS A NEW ARMY OF THE UNDEAD AND I'M TESTING A COUPLE NEW VIRUSES...

YOU DON'T HAVE TO WORRY ABOUT HIM; HE'S NEVER GETTING OUT OF HERE ALIVE.

SLUDGE!

RELAX... I'VE HEARD BECOMING A ZOMBIE IS ONE OF THE MOST PAINFUL WAYS TO DIE.

THAT DOESN'T MAKE ME FEEL ANY BETTER...

I'M A *HERO!*

I'M NOT GONNA LET YOU--

HERO? THERE ARE SUPERVILLAINS WHO DON'T HAVE YOUR KILL COUNT.

WE'LL FIGURE SOMETHING OUT, BUDDY.

FORGET CAVERN KID... LOOK, YOU DON'T HAVE WHAT IT TAKES TO JUMP IN THE DEEP END... OBVIOUSLY.

THIS IS REAL SAD, SIS.

YOU NEED TO START SMALL.

WHY DON'T YOU JOIN THE PEACE CORPS OR SOMETHING...

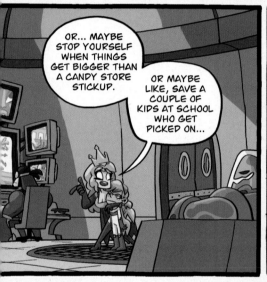

OR... MAYBE STOP YOURSELF WHEN THINGS GET BIGGER THAN A CANDY STORE STICKUP.

OR MAYBE LIKE, SAVE A COUPLE OF KIDS AT SCHOOL WHO GET PICKED ON...

YEAH, WITH ALL OF YOUR POWERS, THINK OF HOW MUCH LUNCH MONEY YOU CAN KEEP OUT OF THE HANDS OF BULLIES.

...

YOU'RE **NOT** HELPING!

COME ON, YOU'RE TALKING ABOUT **BABY** STUFF.

I WANT TO BE A **REAL** HERO!

WITH **FANS!**

AND NEWS COVERAGE!

I JUST...

I JUST DON'T WANT TO SEE YOU FAIL.

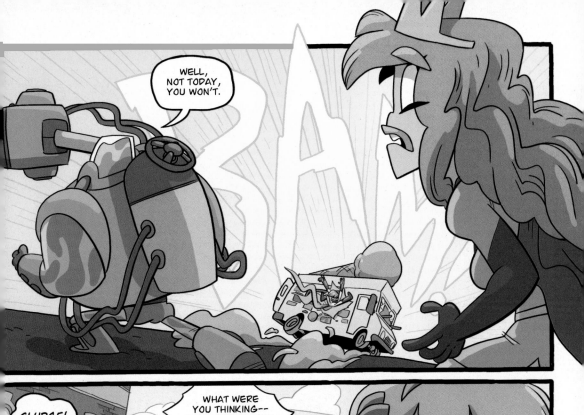

WELL, NOT TODAY, YOU WON'T.

SLUDGE!

WHAT WERE YOU THINKING-- INTERFERING WITH OUR PLANS?!

SLUDGE, STOP!

IT'S--

YOU STOP!

I KNOW WHAT I'M DOING.

JUST...

JUST GO EASY ON HER...

GO EASY ON ME?!

THE ANTIDOTE?!

WHOA, SLOW DOWN, KID!

YOU GOTTA SAVE SOME FOR ALL THE *OTHER* CHILDREN!

MORE!

WITHOUT THE ANTIDOTE THE REST OF THE KIDS WILL TURN INTO ZOMBIES WITHIN THE HOUR...

WELL... TAKE BRODIE TO THE HOSPITAL. AT LEAST *HE'LL* BE ALRIGHT.

WHAT ABOUT THE *OTHER* CHILDREN?

KU-CHUNK

I DON'T KNOW WHAT HAPPENED.

I GOT REALLY MAD BECAUSE THEY WERE TRYING TO HELP ME...

AND NOW SLUDGE IS--

WHAT ARE YOU TALKING ABOUT?

IF ONLY I COULD DO IT ALL OVER AGAIN...

INTERESTING.

I PROMISE I'LL NEVER TRY TO DO ANYTHING GOOD EVER AGAIN!

WHAT THE HELL IS--

--I'LL COME BACK LATER.

AND YOU! FOR THE LOVE OF GOD, STOP BEING BAD... I WON'T BE ABLE TO HELP MYSELF!

DON'T MAKE ME KILL YOU TOO!

WHAT THE FUCK IS SHE TALKING ABOUT?

I'M SORRY! I'M SO, SO SORRY!

I DON'T EVEN KNOW MY OWN STRENGTH!

HE WAS SO GROSS... I MEAN THAT IN A GOOD WAY.

SLUDGE ALWAYS LIKED IT WHEN I CALLED HIM GROSS...

DON'T BEAT YOURSELF UP ABOUT IT TOO MUCH. IT'S GOING TO BE OKAY.

THESE THINGS HAPPEN.

WHAT ARE YOU **TALKING** ABOUT?!

SLUDGE IS **DEAD**!

D-E-A-D! **DEAD**!

HE'S DEAAAADDD!

OH MY **GOD!** WE HAVE TO GO BACK AND GET HIS BODY! WE HAVE TO BURY HIM WITH MOM AND DAD. THAT'S WHAT HE WOULD HAVE WANTED.

?

THAT SOUNDS GOOD FOR YOU, BUT **MAYBE** SOME OF US NEVER **REALLY** LIKED HIM TO BEGIN WITH...

I SAY MAYBE WE LET THAT **LOSER'S** BODY **ROT**.

SHE WAS DOING JUST FINE WITHOUT YOUR CONTRIBUTION, NECROSIS!

GO ON RAE, YOU WERE SAYING HOW UTTERLY DISGUSTING I AM...?

SLUDGE?!

UNUSED ISSUE TWO COVER BY DEREK HUNTER

CHAPTER THREE

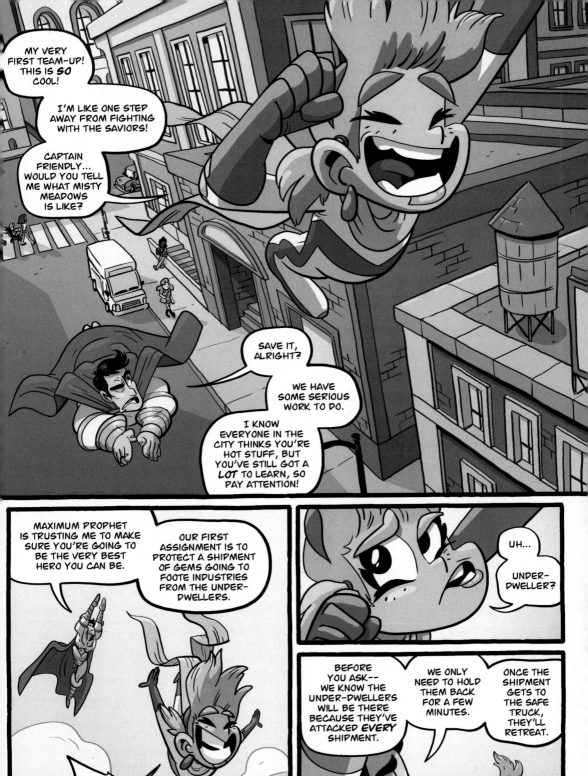

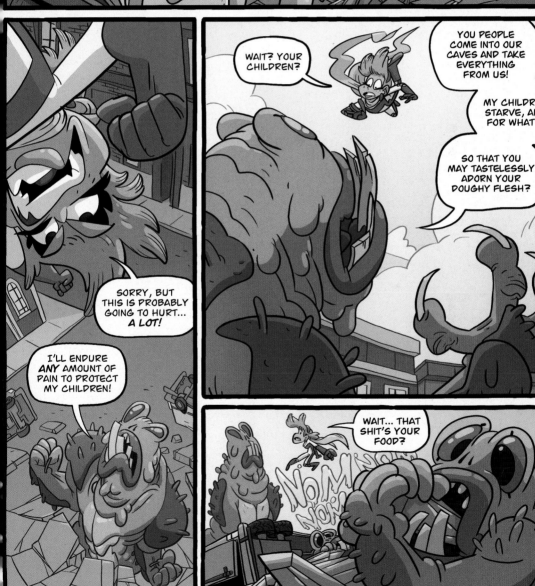

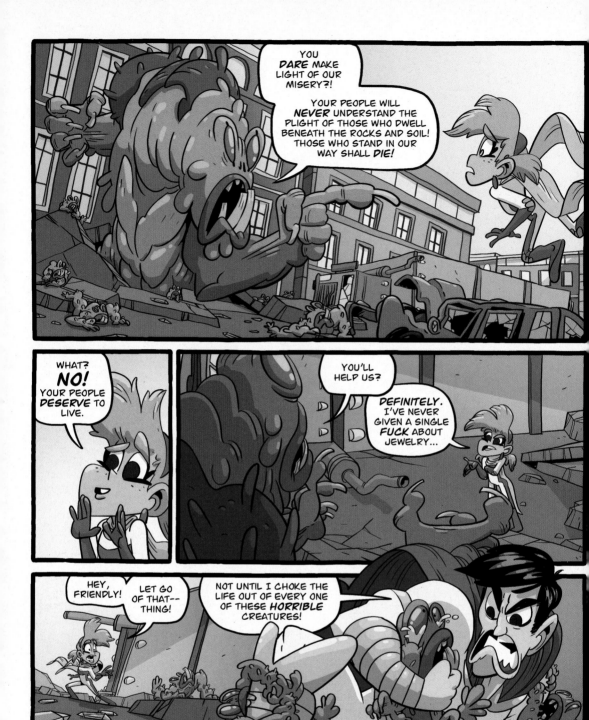

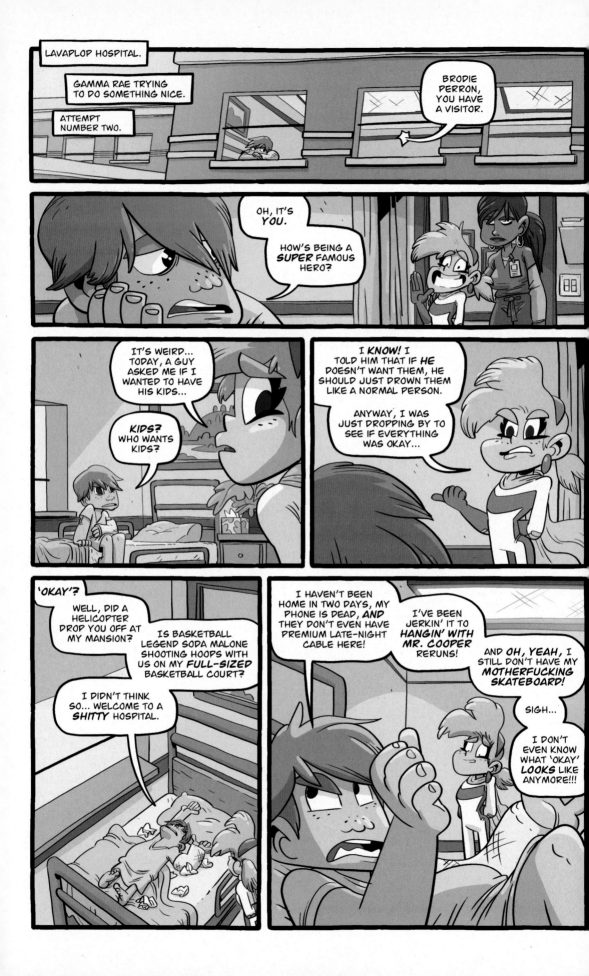

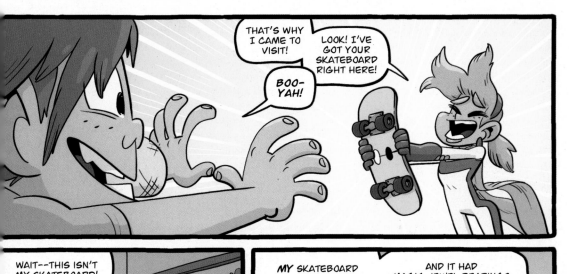

THAT'S WHY I CAME TO VISIT!

LOOK! I'VE GOT YOUR SKATEBOARD RIGHT HERE!

BOO-YAH!

WAIT--THIS ISN'T MY SKATEBOARD!

YOU MUST BE THE WORST HERO OF ALL TIME.

HOW DO YOU KNOW?

MY SKATEBOARD WAS UNBREAKABLE.

AND IT HAD MAGIC JEWEL BEARINGS THAT KEPT IT IN PERPETUAL MOTION. MY DAD PAID A GUY TO BUY IT FOR ME!

I MEANT ABOUT-- UH, NEVER MIND. ARE YOU SURE IT'S NOT YOUR SKATEBOARD?

I FOUND IT RIGHT WHERE YOU SAID YOU LOST IT...

THIS STILL HAS THE TAG ON IT... DO I LOOK LIKE AN IDIOT?

LOOK, I HAD A HARD DAY, ALRIGHT?

I ALREADY TOLD YOU! THAT ZOMBIE LADY STOLE MY BOARD.

NOW DO YOUR JOB AND GET IT BACK!

BACK HOME...

I DON'T HAVE IT ANYMORE.

AND I DON'T KNOW WHERE IT WENT.

WHY DID YOU TAKE IT IN THE FIRST PLACE?!

THAT'S SO ANNOYING!

YOU DON'T EVEN *SKATEBOARD!*

BUT *HE* DOES.

BRODIE PERRON IS A SPOILED SHITHEAD.

LOOK, ALRIGHT, I CAN SEE YOU'RE HAVING A BAD DAY, WHAT'S WRONG?

IT STARTED OUT GOOD, I WAS IN A TEAM-UP.

I HELPED THE UNDER-DWELLERS GET THEIR FOOD BACK--

WELL, *THAT* SOUNDS GOOD, WHAT'S WRONG WITH THAT?

WELL, *APPARENTLY* A LOT, BECAUSE IT'S WORTH LOTS AND *LOTS* OF MONEY...

...AND FOR *SOME* REASON IT DOESN'T REALLY BELONG TO THEM?

AND I GUESS THE MINERS DON'T WANT TO HAVE TO GO INTO THE UNDERGROUND TO MAKE THEM THROW IT UP.

AND MAYBE ALSO, CAPTAIN FRIENDLY *LIKED* HAVING SKIN...?

YOU *KILLED* CAPTAIN FRIENDLY?!

WELP, SEE YOU GUYS.

TAKING CAVERN KID FOR A WALK.

YOU'RE *STILL* KIDNAPPING CAVERN KID!?

IT'S NOT KIDNAPPING! HE LIKES IT HERE.

DON'T YOU, BOY?

SQUEE

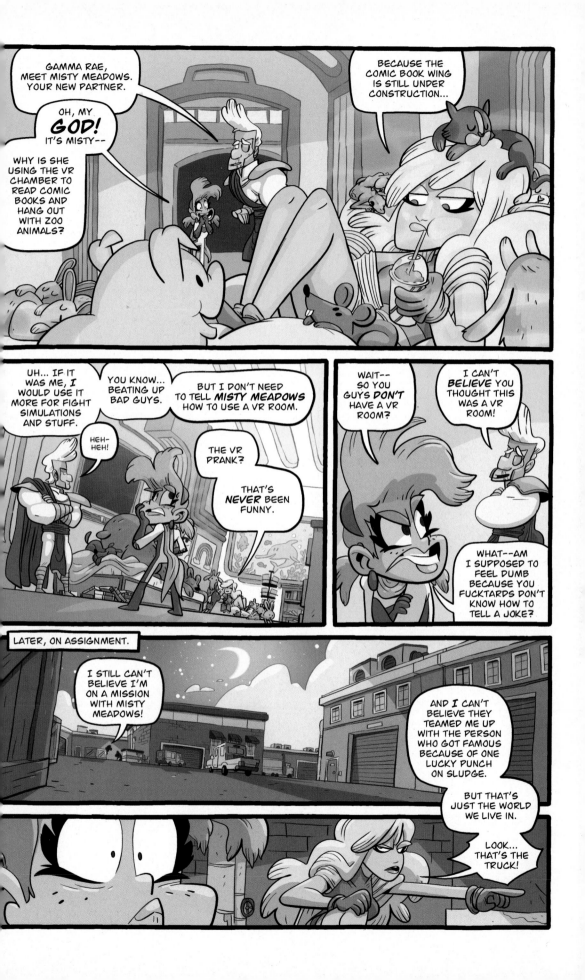

GOTCHA!

AND *THAT'S* HOW IT'S DONE!

SOMETIMES YOU GOTTA USE YOUR HEAD, MY FRIEND.

UH...

I'M SORRY I MESSED UP BACK THERE... IT'S JUST... I WANTED YOU TO LIKE ME.

YOU ACTUALLY DID OKAY.

BUT THAT STILL DOESN'T MAKE US FRIENDS.

I WOULD *NEVER* ACTIVELY BEFRIEND SOMEONE WHO FALLS FOR MAXIMUM PROPHET'S VR CHAMBER GAG.

IT'S A GOOD TRICK!

NO IT ISN'T.

COME ON, LET'S GET INTO THE WAREHOUSE AND INTERCEPT THAT WEAPONS ORDER.

EASY PEASY.

PLEASE LET ME GO! I HAVE KIDS.

YOUR *MOM* HAS KIDS...

AND THEY WOULD BE *VERY* DISAPPOINTED.

ISN'T THAT RIGHT, MISTY?

SURE.

NEVER MIND.

THE IMPORTANT THING IS THAT RIGHT BEHIND THIS DOOR, WE HAVE ALL THE PROOF IN THE *WORLD* TO SEND YOUR BUTT TO *JAIL!*

CHK-CHK CHK-CHK

IT'S... EMPTY.

HOW COULD ALL OF OUR INTEL BE WRONG?

WAIT... THAT SYMBOL ON THE WALL... ISN'T THAT--

YUP... *MERC'S* INSIGNIA.

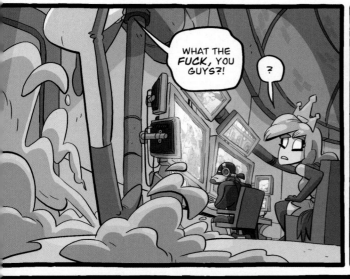

WHAT THE *FUCK,* YOU GUYS?!

?

WHY DO YOU *REFUSE* TO USE THE DOOR?

YOU'RE MAKING IT *IMPOSSIBLE* TO DO MY JOB!

I'M SORRY, RAE... I FEEL BAD... YOU TRUSTED ME, DIDN'T YOU?

DON'T *APOLOGIZE* TO HER!

SHE OWED US FOR THE LAST TIME WE HELPED HER OUT.

BY THE WAY, THANK YOU FOR THE GUNS, SIS.

OKAY, OKAY... YOU'RE RIGHT.

I WOULDN'T EVEN *BE* ON THE TEAM WITHOUT YOU GUYS.

AND AT LEAST I MANAGED TO SAVE THOSE *DISGUSTING* UNDER-DWELLERS...

SO, I'M STILL GOOD. I'M A HERO.

HEY, GUYS!

I THINK I'M STARTING TO *LIKE* THIS CAVERN KID...

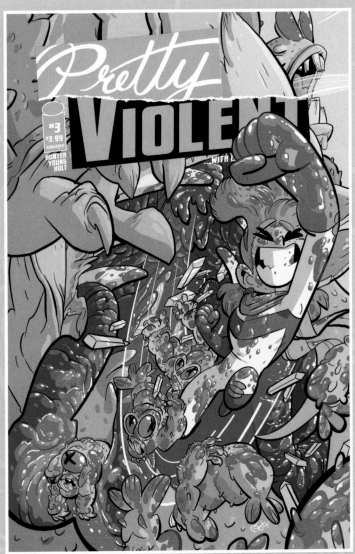

UNUSED ISSUE THREE COVER BY DEREK HUNTER

CHAPTER FOUR

Hey Misty.

I think ur cool.

<3 <3

Who is this?

Gamma Rae!

Got ur number from Max Prof.

LOLZ.

Tuesday 10:20 AM

ur amazing.

Tuesday 9:31 PM

Where do you sleep at night?

The complex?

Wednesday 2:46 AM

nvm... I'll just ask MP.

LOL.

Wednesday 6:02 PM

Want to hang out l8rz? I got some major GRL gossip... :)

No.

Absolutely not.

Thursday 6:03 AM

This is Gamma Rae, btw.

OH, **COME ON**, BE **REAL**. SHE LIKES ME.

I MEAN, **YOU GUYS** TREAT ME LIKE THIS ALL THE **TIME**.

THAT'S BECAUSE **WE'RE** ASSHOLES!

SORRY, RAE. SOMETIMES, WHEN YOU GIVE PEOPLE A CHOICE, THEY DON'T CHOOSE WHAT YOU WANT.

SO... THAT'S IT THEN?

HMM...

MY ONLY FRIENDSHIP IN THE SAVIORS IS RUINED...

NO, WHAT I'M **SAYING** IS, YOU DON'T **GIVE** THEM A CHOICE!

I'VE GOT A TOXIN THAT WILL TURN HER INTO YOUR UNDEAD SLAVE LICKETY-SPLIT!

NO... I JUST WANT HER TO LIKE **ME** LIKE I LIKE **HER**.

AND AS YOUR UNDEAD SLAVE, SHE'LL HAVE NO CHOICE!

HEY, MERC, CAN I COME IN?

YUP! ALL FINISHED!

FOLLOW MY INSTRUCTIONS AND I **GUARANTEE** THAT MISTY MEADOWS WILL LIKE YOU ENOUGH BY THE END OF THE WEEK TO AT **LEAST** RESPOND TO HALF OF YOUR MESSAGES.

WIL--**WHAT?** HOW DID YOU KNOW?

YOU'LL BE HAPPY TO KNOW THAT I HAVE YOUR PHONE TAPPED, AND YOU JUST NEED TO SIGN THERE TO SOLVE ALL OF YOUR PROBLEMS.

OH, THANK **GOD!**

WAIT, WHAT **IS** THIS?

THAT, MY DEAR SISTER, IS A CONTRACT.

IT SIMPLY STATES THAT SINCE I HELPED YOU IN YOUR TIME OF NEED, YOU WILL BE OBLIGATED TO PERFORM ANY ACTION OF MY CHOOSING.

THIS OF COURSE EXCLUDES ANY EXPLICITLY **ILLEGAL** ACTIVITY OR ACTION THAT WOULD PUT YOU IN A POSITION TO DIRECTLY BE THE CAUSE OF HARM TO ANOTHER... AND WHATEVER OTHER **LAME-ASS** HERO ETHIC YOU ASCRIBE TO.

OH, AND THE CONTRACT IS NON-NEGOTIABLE.

OH, NO, I'M NOT FALLING FOR THIS AGAIN! I DON'T WANT ANYTHING TO DO WITH YOUR GUYS' ZOMBIE SLAVES... THAT WAS THE WORST CHRISTMAS OF MY LIFE!

DON'T WORRY, I MADE LOTS OF COPIES.

ALSO, MISTY WON'T **BE A** ZOMBIE.

SHE WILL LIKE YOU.

SHE WILL LIKE **YOU** THE WAY THAT **YOU** LIKE HER.

NOW, GET OUT OF MY ROOM.

WAIT! DO YOU THINK YOU COULD GET BRODIE PERRON TO LIKE ME WHILE YOU'RE AT IT?

NOT POSSIBLE.

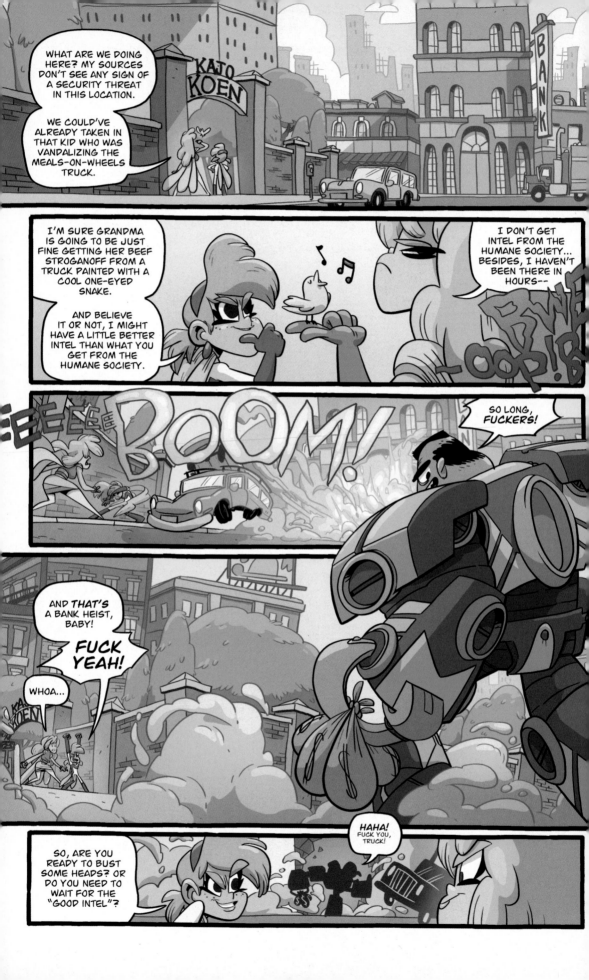

SHIT! SHIT! SHIT! I *HATE* DOGS!

GET *OFFA* ME, MUTTS!

THIS'LL BUY ME A SECOND.

GRRR.

Grr...

SO LONG, POOCHIES!

SNOOPY! FARFADIL!

TUUUUU-PAAC!

KA-BOOOM

MY DOGS...

NO TIME TO CRY OVER SPILT DOG ENTRAILS, MISTY MEADOWS.

THIS ASSHOLE'S PUPPY-MURDERING DAYS ARE *OVER!*

BACK HOME.

I'M A *LITTLE* SAD THAT ALL OF MISTY'S DOGS DIED, BUT WE *WERE* ABLE TO RECOVER THE MONEY HE TOOK FROM THE BANK.

THIS NEXT ONE IS GOING TO BE EVEN EASIER.

THIS GUY LOOKS TOUGH, BUT HE MELTS WHEN HE SEES KITTEN MEMES.

BELIEVE ME.

THIS IS *A LOT* OF DIRT ON *A LOT* OF BAD GUYS... HOW DO YOU KNOW ALL THIS?

SEARCH HISTORIES. MEDICAL RECORDS. PSYCHIATRY REPORTS. EVERY INDIVIDUAL'S WEAKNESS IS OUT THERE. YOU JUST HAVE TO KNOW WHERE TO LOOK.

SO, YOU'RE BASICALLY AN *EVIL* LIBRARIAN?

I ALWAYS THOUGHT YOU WERE, LIKE, A MASTER ASSASSIN OR SOMETHING COOL.

YOU'D DO WELL TO *RESPECT* ME, LITTLE SISTER.

WHAT ARE YOU GONNA DO? TELL THE WORLD I'M MILDLY ALLERGIC TO GUINEA PIGS?

OH, *NO!* I'D BE *DONE* FOR!

PLEASE, MERC, IF THIS INFORMATION GETS INTO THE *WRONG HANDS*... I *MIGHT* BE FORCED TO TAKE AN *ANTIHISTAMINE* BEFORE WORK!

ZOO??

OK, *JEEZE.* SORRY.

JUST DO WHAT THOSE FILES SAY, AND THEN FULFILL *YOUR* HALF OF THE BARGAIN.

THESE FILES ARE THE BEST THING THAT'S EVER **HAPPENED** TO ME.

I LOVE EACH AND EVERY ONE OF YOU.

DON'T FORGET, RAE, THAT INTEL COMES AT A PRICE.

YEAH, YEAH, WHATEVER. I'M GOING TO MEET MISTY AT SAVIOR COMPLEX. CAN YOU JUST E-MAIL ME THE NEXT BATCH?

BYEBYEBYE!

SAVIOR COMPLEX.

OUT OF MY WAY, LOSERS! COMING THROUGH!

HEY! **TRAIN GUY!**

WHAT ARE **YOU** DOING HERE?!

OH, GODDAMMIT...

LEMME GUESS... BECAUSE I'M IN A **WHEELCHAIR** THAT MEANS I CAN'T BE A SUPERHERO?

YEAH. PROBABLY NOT...?

BUT THAT'S NOT WHAT I MEANT--JUST THAT SECURITY'S PRETTY TIGHT AROUND HERE. I ONCE TRIED TO SMUGGLE A GALLON OF NITROGLYCER--

OH, SO NOW YOU'RE IMPLYING **I** CAN'T GET THROUGH SECURITY WITH NITROGLYCERIN?

I'M SAYING **NOBODY** CAN. **EVEN** IF THEY'RE CAPABLE.

BRAINDEAD ABLEIST!

WHAT IN THE HELL IS GOING ON HERE?!

HE STARTED IT!

WHEN YOU ASSUMED I WANTED TO KILL MYSELF BECAUSE I CAN'T FLY!

NO, BECAUSE YOU CAN'T WALK!

YOU ASS!

GUYS, GUYS, GUYS! STOP!

GAMMA RAE! THIS IS MY BROTHER, PAUL.

OKAY.

COOL.

WELL, ANYWAY, LET'S GO, MISTY. THOSE FROZEN YOGURTS AREN'T GOING TO EAT THEMSELVES... ALTHOUGH, I'M SURE THEY WOULD LIKE TO.

ARE YOU FOR REAL?

I'M NOT GONNA GO HANG OUT WITH YOU!

NOT NOW!

I'M GONNA STAY AND HANG OUT WITH MY BROTHER. JEEZE...

WHAT?

THAT'S OKAY, I WANTED TO HANG OUT ON MY OWN ANYWAY!

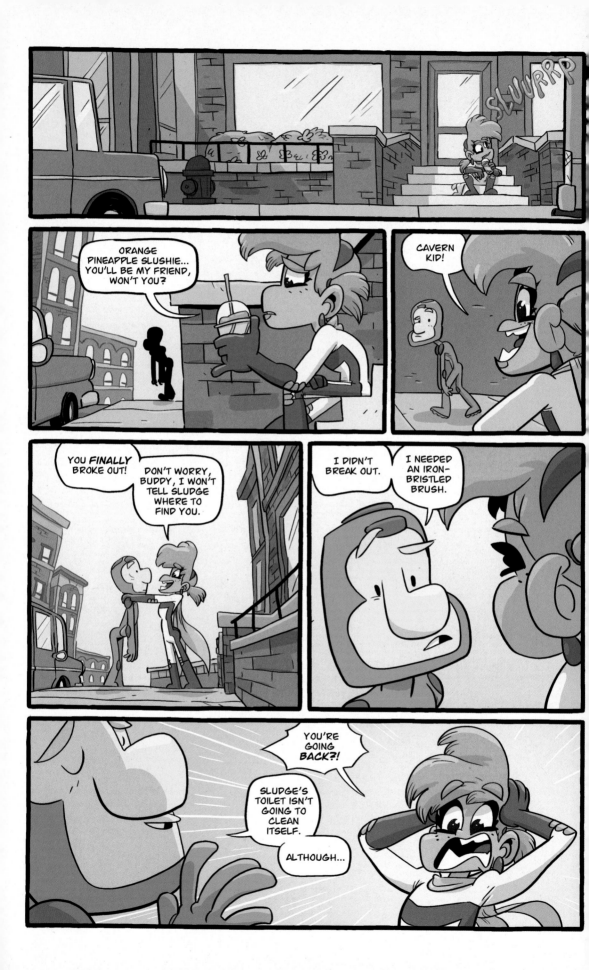

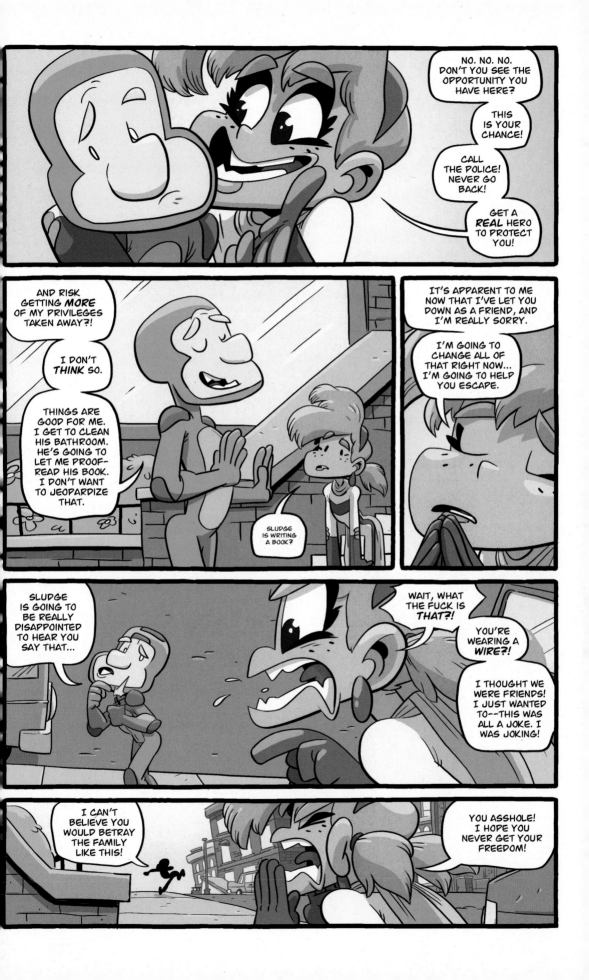

BAY CITY HOSPITAL.

BRODIE?

YOU *JUST* MISSED HIM... HE WAS DISCHARGED A LITTLE WHILE AGO.

I KNOW IT'S NONE OF MY BUSINESS, BUT IF YOU'RE THINKING ABOUT GOING TO SEE HIM--FROM ONE WOMAN TO ANOTHER... I WOULD ADVISE STAYING FAR, *FAR* AWAY FROM THAT BOY...

NICE TRY, LADY. BUT I'VE PRETTY MUCH HELD BRODIE PERRON'S HAND ALREADY.

HE'S *MINE.*

BRODIE! I'M SO GLAD YOU'RE BETTER. I'VE BEEN WORRIED ABOUT YOU.

WELL... I'M *MOSTLY* BETTER.

I STILL HAVE THE SHITS. OR, *DIARRHEA* AS THESE HOSPITAL SERVANTS CALL IT.

YOU'RE SO FUNNY.

SO... HAVE YOU SEEN ME ON THE NEWS LATELY?

MORE ABOUT ME, LESS ABOUT YOU...

I'M THROWING A BIG BIRTHDAY BASH AT THE ZOO... IF YOU'RE NOT TOO FAMOUS BY THEN, I'D LOVE IT IF YOU CAME AS MY SPECIAL GUEST.

WOW, REALLY? I'D LOVE TO!

OH, AND DON'T FORGET TO WEAR THIS.

"BRODIE... BORN 2 B BAD..."

AWESOME! I'LL BE THERE-- WAIT, ISN'T YOUR BIRTHDAY IN DECEMBER?

IT'S MY HALF BIRTHD-- HOW DO YOU KNOW WHEN MY BIRTHDAY IS?

ARE YOU SOME KIND OF STALKER?

I MEAN... WHAT IS A STALKER ANYWAY?

A PERSON WITH AN UN-HEALTHY OBSESSION WITH KNOWING WHERE YOU ARE AND WHAT YOU'RE DOING ALL THE TIME?

BY THAT DEFINITION, WHO ISN'T A STALKER? COME ON...

SIR, I WOULD HIGHLY ADVISE AGAINST INVITING THIS GIRL TO--

SHUT THE FUCK UP, BENTLEY!

SO DO YOU HAVE ALL THE COOL SECRET CAMERAS AND MICS AND SHIT?!

HELL YEAH!

BADASS.

SEE YOU AT THE ZOO!

BZZT BZZT

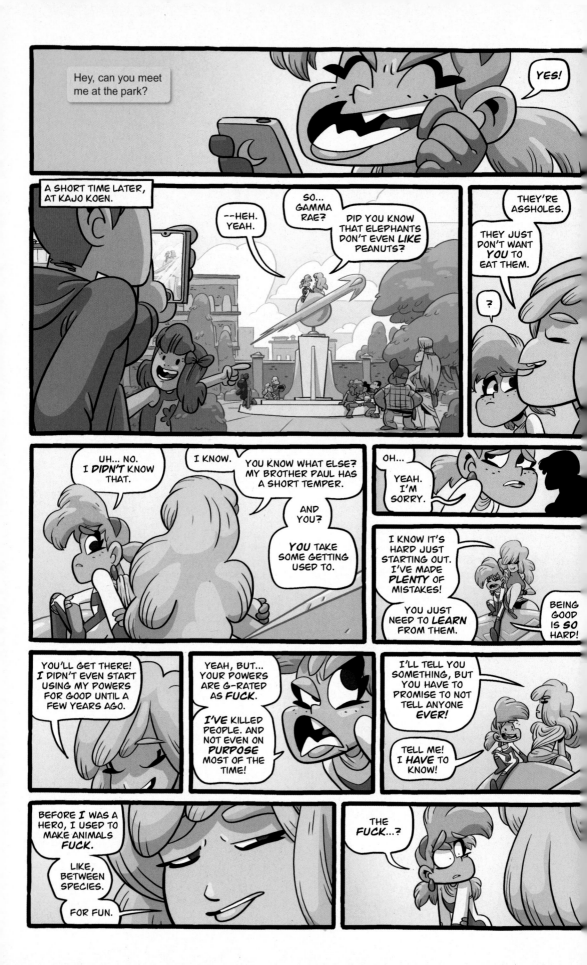

ISSUE ONE VARIANT COVER BY SHANE HILLMAN

CHAPTER FIVE

I MESSED UP, OKAY?!

I'LL MAKE IT UP TO YOU!

IT'S *NOT* OKAY!

YOU'RE SUPPOSED TO BE *BETTER*!

YOU'RE *SUPPOSED* TO ACT LIKE A *HERO*!

WOOH

BUT...

...I DO HAVE A NEW ASSIGNMENT FOR YOU, GAMMA RAE.

OH, GOD, *THANK YOU!* I WON'T SCREW IT UP!

PRO-TEEN IS ASKING FOR ASSISTANCE FEEDING THE HOMELESS...

FEEDING THE HOMELESS?

...TO WHAT?

I'M THE LAUGHINGSTOCK OF THE *ENTIRE* SUPERHERO COMMUNITY!

AND EVERYONE THINKS MY WEAKNESS IS A *GODDAMN* CHILDREN'S TOY!

THEY'VE GOT ME FEEDING *GRUEL* TO PEASANTS!

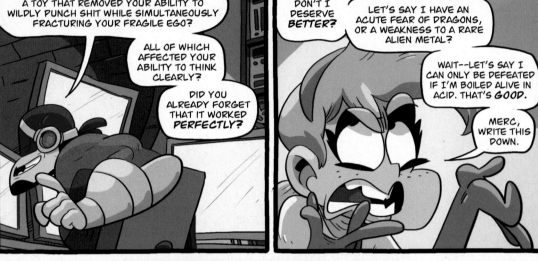

A TOY THAT REMOVED YOUR ABILITY TO WILDLY PUNCH SHIT WHILE SIMULTANEOUSLY FRACTURING YOUR FRAGILE EGO?

ALL OF WHICH AFFECTED YOUR ABILITY TO THINK CLEARLY?

DID YOU ALREADY FORGET THAT IT WORKED *PERFECTLY*?

DON'T I DESERVE *BETTER*?

LET'S SAY I HAVE AN ACUTE FEAR OF DRAGONS, OR A WEAKNESS TO A RARE ALIEN METAL?

WAIT--LET'S SAY I CAN ONLY BE DEFEATED IF I'M BOILED ALIVE IN ACID. THAT'S *GOOD*.

MERC, WRITE THIS DOWN.

RAE... THAT'S YOUR *EGO* TALKING AGAIN.

YOU WERE *SUPPOSED* TO LOSE TO KILL COUNT, THAT WAS OUR DEAL.

THAT WIMPY *ASSHOLE* CALLS HIMSELF *KILL COUNT*?!

RAE, DON'T START. SLUDGE AND NECROSIS WOULD AGREE WITH ME ON THIS...

STAY AWAY FROM HIM!

I *REALLY* WISH I COULD HELP YOU OUT.

YOU KNOW, BECAUSE WE'RE SIBLINGS AND WE *LOVE* EACH OTHER.

BUT *LEGALLY* I'M NOT REALLY OBLIGATED TO DO WHAT YOU SAY...

YOU KNOW... NOT WITHOUT A CONTRACT...

PRICK...

TOP SECRET

BACK AT THE SAVIOR COMPLEX.

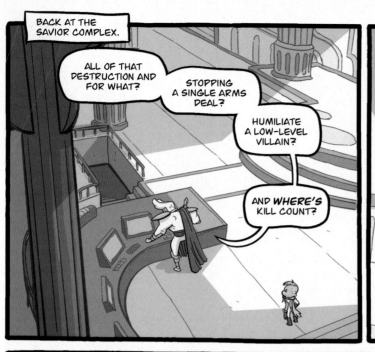

ALL OF THAT DESTRUCTION AND FOR WHAT?

STOPPING A SINGLE ARMS DEAL?

HUMILIATE A LOW-LEVEL VILLAIN?

AND **WHERE'S** KILL COUNT?

I DON'T KNOW. I'M JUST TRYING TO DO SOME GOOD-- LIKE YOU SAID!

WELL, THIS **ISN'T** GOOD, GAMMA RAE.

IT TURNS OUT KILL COUNT WAS **MASSIVELY** CONNECTED...

IF YOU KNOW WHERE HE IS, AND YOU'RE LYING TO ME... THINGS CAN GET VERY VERY BAD FOR US...

HEY, YOU DON'T EVEN KNOW IT WAS **ME**.

YOU LEFT YOUR **INITIALS** ON THE WALL!

MAYBE IT WAS **GRAMMAR** RAE?

LOOK, I DON'T KNOW **WHERE** HE IS, OKAY?

KILL COUNT...

...WAS HE THERE?

DID YOU KILL HIM?

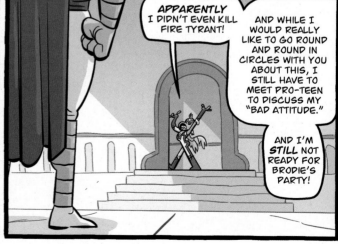

APPARENTLY I DIDN'T EVEN KILL FIRE TYRANT!

AND WHILE I WOULD REALLY LIKE TO GO ROUND AND ROUND IN CIRCLES WITH YOU ABOUT THIS, I STILL HAVE TO MEET PRO-TEEN TO DISCUSS MY "BAD ATTITUDE."

AND I'M **STILL** NOT READY FOR BRODIE'S PARTY!

CAN'T DO ANYTHING RIGHT-- OUTTA MY WAY!

EH...?

HEY! PAUL! CAN YOU TELL MISTY TO COME TO BRODIE'S PARTY?

SHE HASN'T RESPONDED TO MY TEXTS IN DAYS.

SHE CHANGED HER NUMBER. SHE DOESN'T WANT TO TALK TO YOU. *I* DON'T WANT TO TALK TO YOU...

LISTEN... I KNOW I'VE COME ACROSS LIKE AN ASSHOLE TO YOU IN THE PAST, AND I WANTED TO SINCERELY APOLOGIZE, AND TELL YOU THAT, WELL--YOU'RE RIGHT. ANYONE CAN BE A HERO.

*SUPER*HERO.

...

ARE YOU GOING TO TELL MISTY TO COME OR NOT?

WE WEREN'T EVEN INVITED.

I'M THE SPECIAL GUEST, YOU TWO ARE *DEFINITELY* INVITED.

BESIDES, IT'S NOT LIKE ANYONE WOULD KICK SOMEONE OUT WHO--NEVER MIND.

"SOMEONE WHO" WHAT?

SAY IT!

GOTTA GO!

SEE YOU THERE!

IDIOT.

NECROSIS?

MERC?

SLUDGE?

...

CAVERN KID?

LOOKS LIKE I'VE GOT THE PLACE ALL TO MYSELF FOR ONCE.

SMART?

CHECK.

BEAUTIFUL?

DOUBLE CHECK.

PLUS... THE BEST PRESENT ANYONE HAS EVER SEEN...

TODAY IS THE DAY YOU *FINALLY* SEPARATE YOURSELF FROM THE REST OF THE HERD, YOUNG LADY.

ISSUE SIX PINUP BY JAKE WYATT

CHAPTER SIX

BUT WHAT WAS SHE *DOING* THERE?!

LISTEN, THE POINT IS, IF MY SON SHOWS UP WITH SO MUCH AS A *BRUISE*, I WILL SPARE NO EXPENSE MAKING YOU REGRET *EVERY SINGLE SOLITARY* MOMENT OF THE REST OF YOUR *SAD, MISERABLE, PATHETIC SO-CALLED LIFE!*

NO ONE'S HEARD FROM HIM SINCE YOUR SISTER INFILTRATED HIS HIDEOUT.

DO YOU THINK YOUR SISTER IS GOING TO BE ABLE TO KEEP BEING A SAVIOR WHEN THE PUBLIC FINDS OUT WHO SHE *REALLY* IS?

WE'RE *GOING* TO FULFILL OUR HALF OF THE CONTRACT, BUT ONCE WE GET KILL COUNT BACK--IT'S OVER...YOUR BLACKMAIL WON'T WORK AFTER THAT.

THERE'S NO SIGN THAT HE WAS EVEN THERE WHEN SHE STRUCK--

NO MATTER HOW MUCH OF A TOUGH ACT YOU PUT UP, YOU AND I BOTH KNOW THAT YOU'D *NEVER* ACTUALLY LET ANYTHING HAPPEN TO YOUR LITTLE SISTER...

...WOULD YOU?

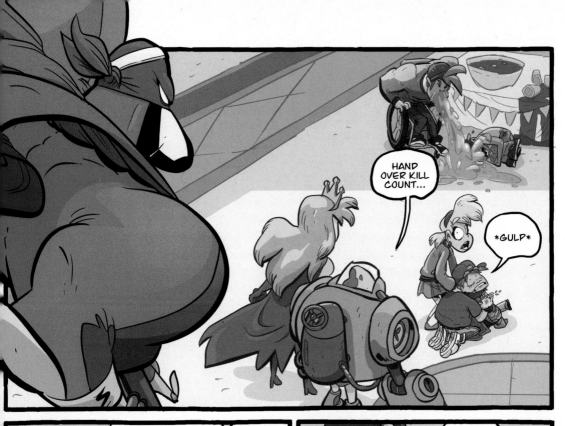

HAND OVER KILL COUNT...

GULP

YOU'LL HAVE TO KILL HER TO GET HIM!

HEY!

AND *THEN* ALL OF MY BUTLERS!

SOUNDS FUN.

SHALL WE?

AND ME!

COWARDS.

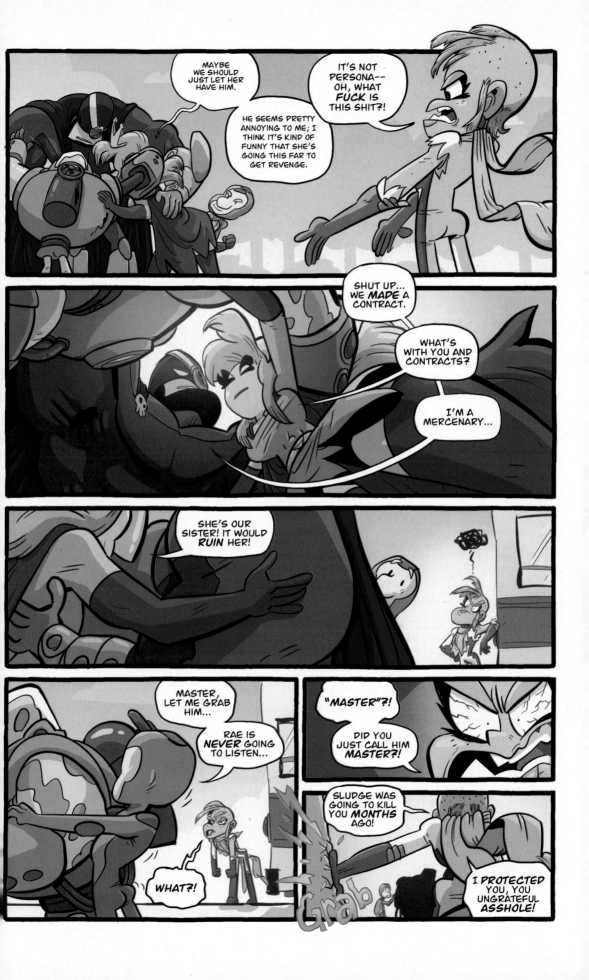

YOU SHOULD BE CALLING *ME* MASTER!

ME!

WELP, THERE GOES DOING THINGS THE EASY WAY...

NOW, RELEASE KILL COUNT IMMEDIATELY...

AW... IT'S OKAY, SWEETIE, YOU'VE TRIED YOUR BEST.

HUMP! HUMP! HUMP!

...BUT I'LL TAKE CARE OF THIS ONE *MYSELF!*

FUCK.

WHAT THE *FUCK?!*

WHAT'S HAPPENING?!

MOMMY!

HUMP HUMP HUMP

NO.

NO!

DON'T GIVE IN! LIKE ALL MERE MORTALS, THEY'LL EVENTUALLY RUN OUT OF STAMINA!

FUCK!

MERC!

TAKE ONE MORE STEP TOWARDS KILL COUNT AND MISTY... YOU'LL REGRET IT.

COME ON, LIKE I'M GOING TO START TAKING ORDERS FROM YOU.

YOU KNOW, I WAS YOUR AGE WHEN THEY LEFT US...

...MOM AND DAD.

I WASN'T READY TO TAKE CARE OF A WHOLE FAMILY, BUT I STEPPED UP, BECAUSE THERE WAS NO ONE ELSE--

NO! IT'S OVER NOW, SO YOU CAN SAVE YOUR SHITTY MONOLOGUE FOR SOMEONE WHO GIVES A SHIT!

HEH.

STOP! I SAID DON'T MOVE!

OH, MAN...

NO!

GAMMA RAE...?

I'M HERE.

I'M HERE FOR YOU.

I SHIT MY PANTS.

TELL MY BUTLER TO... CHANGE MY UNDERWEAR.

BRODIE, YOU *CAN'T* DIE!

GAMMA RAE...?

HE IS *MY* BOYFRIEND, I'D LIKE TO--

I HATE YOU!

I HATE YOU ALL!

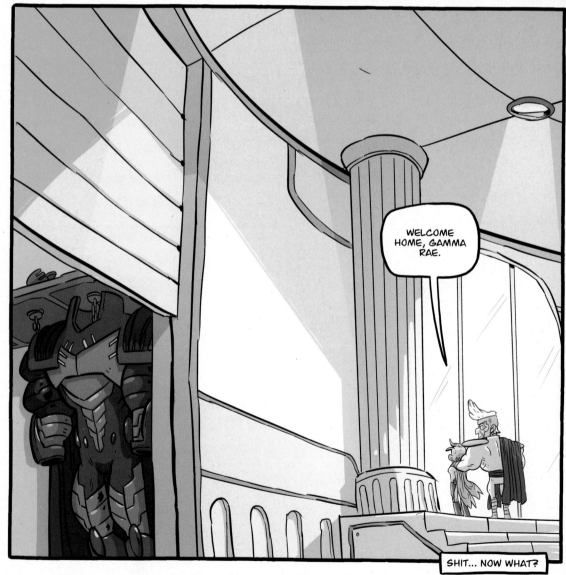

THE END

VOLUME ONE

BONUS COMICS

ISSUE ONE BONUS
DEREK HUNTER & JASON YOUNG

ISSUE FIVE BONUS
ROSE HUNTER

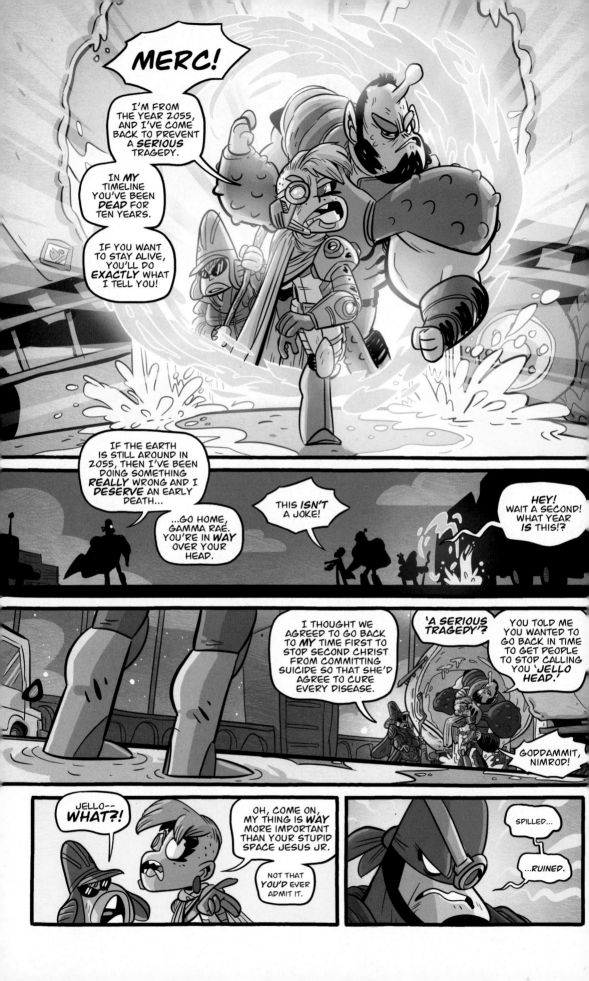

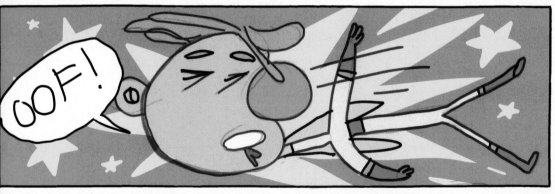

PRETTY VIOLENT ORIGINS

Pretty Violent, in its original form, was based on me and my life growing up as a leftist, punk rock, art weirdo in a very conservative and extremely religious household. And like Gamma Rae, I had a family that loved and supported me in my social and creative pursuits despite our differences. That's how it all started, wanting to tell a personal story about not fitting in. About having one foot in one world, one foot in another, and as a result, never really feeling like I found a place for myself in either one.

Once the framework of the character relationships were developed, I started designing everyone, and, well... the plucky, poofy-haired redhead you see below, the original protagonist of my future story, wasn't doing it for me. So, I looked at my own little family and saw a lot of fun possibilities for character traits in my then 5-year-old daughter. A sweet kid who tried so hard to be good, and found herself frustrated to tears (and sometimes fits of rage) when she didn't live up to her own expectations for herself.

What a perfect little trait to draw from for a story about a superhero raised by villains. The main character had to be based off my own daughter.

It was perfect.

Clockwise from top: Original sketches of Kill Count & "Gamma Rae", initial line-up of characters for the first arc, original cover with the old title, early Madmanimal sketch.

As you can see, three characters were completely left out, including the first final drawing of "Little Hero" (Our placeholder title for the book at the time).

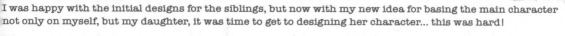

I was happy with the initial designs for the siblings, but now with my new idea for basing the main character not only on myself, but my daughter, it was time to get to designing her character... this was hard!

Three different designs, none of which excited me very much... not for the main character at least. I'll probably end up using the bottom design for something... I like that one!

This image was drawn as part of the first pitch I was going to send off to Image... the main character design, while better than the others, was still not right!

Speaking of "Still Not Right", below are some unused designs for characters that appeared in the first six issues.

Early explorations for a character who "dies" in issue 1. She was too cool to waste, so she can be seen "on the mend" later in the series. She'll be back.

This early **Hulktress** design was a little too "on the nose" even for me.

An early design for **Fire Tyrant** from issue 4. I liked it enough to tweak the design to use for the "Unknown Villain" in issue 6.

Title page from the **Gamma Rae** short story in **Popgun: Vol. 1**

Cover for the **Gamma Rae** board book I made for my daughter Rose shortly after she was born.

After each failed attempt designing a new costume for the hero of our book, my mind kept going back to the design for **Gamma Rae**, a character I designed for a short story I'd written for Image Comics' **Popgun Anthology**; I kept wishing I could design something as cool as that for this book.

Then I realized it was dumb to waste a perfectly good design for a character that I'd barely even used.

MAKING A PAGE

1: Roughs are drawn (often during the scripting phase) to get a quick idea of how much information can be stuffed into a panel.
2: Pencils tighten up the mess, background characters are fleshed out (this is also when I lay out the final lettering).
3: Final inks are done on 2 layers, a character and background layer, to make separation of different elements easier for Spencer.

As the series continued and I got more comfortable drawing the characters, my "pencils" got a lot looser... to the point where most people wouldn't call them pencils at all. Here's a panel from issue 6 for comparison to my issue 1 process above.

Derek Hunter is a Los Angeles-based animation designer known for his work on shows like **DuckTales** and **Adventure Time**. He is less known for his previous comic book series, Pirate Club.

He's been avoiding a healthy social life in favor of spending nights and weekends making comics since he was 10.

Special Thanks to Rachel and my kids for being so supportive as I tackled the challenge of making a comic book series while juggling a full-time job. **Many thanks** to Ryan Ottley, Tanner Johnson, Jake Wyatt, Matt Youngberg, Skottie Young, Robert Kirkman, Cory Walker, Andy Ristaino, Shane Hillman, Frank Angones, and all of my friends who have been so encouraging and supportive throughout this whole process.

Jason Young has been writing comics and zines since 2008.

He was a developer on the Image comic **I HATE FAIRYLAND**. **PRETTY VIOLENT** is his fourth writing collaboration with his frequent creative partner, Derek Hunter. He lives in Salt Lake City with his cat, Puss-Hole.

He begs you on hands and knees to follow him on social media so you can hear about all of the projects that are coming on the horizon. **@theejasonyoung**

Spencer Holt has been making comics for fun and profit for most of his life, contributing color work on books like **GREEN MONK**, **Jim Henson's Storyteller: Giants**, and **PRETTY VIOLENT**.